Y0-CAF-989

How to Grow Food: A Wartime Guide was written in 1940 by the well-known novelist Doreen Wallace, as part of the Batsford *Home Front Handbook* series. It was published to help those who moved to the countryside during the war years and needed to learn to grow their own food. The author was not only a novelist and critic but a farmer, and she combined her literary talents with her practical farming knowledge to produce this wartime classic. Warm and witty as well as informative, this is still a great guide for any beginner or experienced gardener.

The author explains exactly what to grow, and when, how and why; she deals with the problems of weeds and pests and their remedies; she covers 'the vexed question of small fruit'; arrangement and rotation; and keeping hens and pigs. She has added an admirable 'ABC of Useful Vegetables'.

This wartime classic is a must for anyone keen to grow their own food.

A BATSFORD BOOK

HOW TO GROW FOOD

First published February 1940

Reprinted 2012

By Batsford
10 Southcombe Street, London W14 0RA

An imprint of Anova Books Company

ISBN: 9781849940498

A CIP catalogue record for this book is available
from the British Library.

18 17 16 15 14 13 12 11 10 9 8 7 6 5 4 3 2 1

Printed and bound by Toppan Leefung Printing Ltd, China
Reproduction by Mission Productions Ltd, Hong Kong

This book can be ordered direct from the publisher at the website:
www.anovabooks.com

I   A WEATHER-BEATEN LADY

DOREEN WALLACE

HOW TO
GROW FOOD

LONDON
B. T. BATSFORD LTD
15 NORTH AUDLEY ST., W.1
and at Malvern Wells

PUBLISHER'S NOTE

The contents of this book is a faithful facsimile of the original 1940 text. As such, it includes the phraseology of the time, the food and farming culture of the time, and a few gardening and farming suggestions that may not be suitable today. Several brand names are no longer available and some organizations have disappeared. In particular, we would like to draw your attention to the following:

Page 36:   Cyanide is not suitable for use when dealing with wasp nests nor is soaking rags in paraffin and setting alight. If there is a serious wasp problem, readers are advised to contact a pest specialist.

Pages 39-49: Various suggestions for pest control may seem quite cruel to today's audience, but the chapter reflects the country views on pest control in 1940. Also, many of the suggested brands and materials are no longer readily available.

Page 45:   Please do take care in any use of a block of meta as it is poisonous if ingested.

Page 46–47:   It is now illegal to take the eggs of any bird or to disturb them in any way.

Page 62:   Paraffin can burn the skin of hens, so it is not advisable for use now.

Page 68:   For 'salad oil' use olive oil.

Page 68:   'B.W.D.' is also known as 'blue wing disease'.

Page 72:   'middlings' refers to wheat middlings

Page 72–76:   All pigs are now subject to laws about what can be fed to them in order to prevent the spread of disease. The suggestions for feeding pigs in these pages are now out of date. If you are planning to keep pigs, please do seek full guidance on feeding and housing pigs.

Page 74:   'Erysipelas' is an acute streptococcus bacterial infection.

Page 82:   'Artichoke, Tubers' refers to Jerusalem Artichokes.

Page 88:   'Vegetable Marrow' also includes courgettes.

CONTENTS

ACKNOWLEDGMENT

The Publishers must acknowledge their obligation to the Photographers whose work is reproduced in the following pages, namely: W. Abbing, for Figs. 42, 46, 49–51; F. C. Brown, for Figs. 6, 16–19, 31–34; Messrs. Dorien Leigh, Ltd., for Fig. 1; Messrs. Hortiphot, for Figs. 8, 14, 15, 24; Messrs. Reginald A. Malby, Ltd., for Figs. 3, 4, 5, 7, 9, 10–13, 20–23, 26, 27, 29, 30, 35–38, 43–45; and Messrs. Musto, for Figs. 2, 28. They are particularly indebted to the Editor of *The Feathered World* for generously loaning the originals of Figs. 39–41; and to Messrs. Carter's Tested Seeds, Ltd., for Figs. 25, 47, 48, 52, whose produce these photographs illustrate. The drawings of gardening implements have been specially made for the book by Miss Norah Davenport, and the line illustrations to the chapter on Weeds and Pests were made by the Author.

2   SECOND EARTHING OF POTATOES

3   LIFTING POTATOES

4 BEAUTY AND UTILITY

I

PREAMBLES AND PREPARATIONS

This little book is not intended to teach the farmer or market-gardener his business. It is for those who have never grown food before, either because they have had no gardens, or because, being possessed of gardens, they have grown nothing but flowers.

To-day we are all urged to grow food, and there are thousands of people who fortuitously find themselves in a position to respond to the urge, though they are without the necessary experience. These are they who have lately moved from town to country and perhaps for the first time in their lives have become garden-owners. An even larger body is that of pleasure-gardeners, both in the suburbs and in the country proper, who would now wish to become useful gardeners. And there is a third body which the Government hopes will arise Phoenix-like from its ashes—the body of allotment holders. When living was dear, after the last war, allotments were popular, but of late they have been almost completely neglected. In our village, nobody will trouble to cultivate an allotment except the man with an outsize in families: he is the only man who feels the pinch of poverty severely enough to try to alleviate it by working after his statutory work-hours.

Now that the country is at war, however, poverty is not the point. Nourishment is the point. Whether one has the money or not, it is possible that on some days of the week one will not be able to buy enough food to fill the inner man. But with a garden of vegetables, the inner man need never feel empty. With a garden of vegetables, one will be able to get over the difficulties of distribution that seem to afflict the Government's well-laid plans for feeding the multitudes. With a garden of vegetables, the country dweller need not toil out on a bicycle in bad weather to buy provisions at the nearest shops, which may be four, five or six miles away. . . .

I am profoundly sorry for townspeople who are now tasting the joys and sorrows of country life for the first time, or who have only tasted them before in holiday-times, when everything was fun and it was all over in a fortnight anyway. To find themselves rooted in the country for the winter as well as the summer, and at a time of petrol-restriction when not only is joy-riding impossible, but the fetching and carrying of supplies wellnigh impossible too, must be something in the nature of a spiritual shock. The inconvenience, the isolation, the cold—it is always colder in the country than in a town—and the monotony of all these things! Myself, I am a countrywoman born and bred, but even I have looked across the hearth, in these long dark evenings, at the sharer (by law) of my joys and sorrows, and have reflected that in this war an appreciable number of casualties will be murders on the home front.

To ease the tension of too constant propinquity, to counteract "states of mind" from whatever cause, there is nothing like gardening. It matters not whether the object of one's labours is a parsnip or a penstemon, the work's the thing. Not every kind of work is efficacious in this way; housework only makes the miserable more miserable. But to work in the open air, at tasks which really need attention, is to diffuse thought and lull it, and at the same time to gain in physical well-being. It is impossible to be consistently unhappy while digging, planting or weeding.

So not only will the vegetable-grower be fulfilling the hopes of the Government and the needs of his own interior: he will be giving himself contentment and steadiness of mind. Personally, I cannot imagine how I could endure country life from week's end to week's end all the year round, though I was brought up to nothing else, without my garden; and I am certain that those townspeople who have migrated to the country will find everything easier to bear if they make the most of their gardens: while as to suburban dwellers, they always *have* made the most of their gardens—a more exemplary set of horticulturists does not exist—but they will enjoy the glow of added virtue if they turn part of their land to food-production in the present crisis.

I hope I shall be forgiven if I address this book chiefly to the utter amateur. He is, after all, more in need of help than those who already know something about gardening. Let us imagine that a large firm has "evacuated" from London to the country. (I do so dislike this medicinal terminology of evacuations and purges, but I have to use the same words as other people if I am to be understood.) A mausoleum of a country mansion houses a good deal of the staff, but Mr. B., an accountant, has been fortunate enough to secure one of the lodges, a mile away across the park, in which to hide Mrs. B. and three little B.s from the fury of the Hun. Formerly, the B.s lived in a flat; they have never had a blade of grass to call their own.

They have now far too many blades of grass, for the place has been empty for some time. There are signs that a large patch of weedy ground at the back of the house, some quarter of an acre in extent, has been under cultivation, and the B.s, who were evacuated (I'm sorry for them, but I can't avoid the word) in the glorious weather of early September, immediately vowed themselves to its reclamation. They would grow food. Vegetables fresh from the soil would be so good for the children, even if the children didn't know it and refused to believe it. "A-a-h!" breathed Mr. and Mrs. B., largely, drinking in the balmy air. How lucky they were in having been able to arrange a country holiday "for the duration."

Now in midwinter they feel a little differently about it. But let us not anticipate. Let us watch them at their childish preparations.

There is not a gardener to be had, for all the men in the tiny village work on farms, and there is no call for jobbing gardeners. Up at the tomb-like mansion, where the gardens are on a vast scale, there is a greatly depleted garden staff consisting of one old man and one young boy, who will have their work cut out to keep down the weeds; the B.s can expect no help there. But the lack of a gardener is the least of worries—they would far rather do all the work themselves anyway. So healthful, so interesting.

Mr. B. takes a trip on his bicycle to the market town to buy tools. He buys them large and heavy, believing all he is told by the shopkeeper. Up to a point, the shopkeeper is to be believed, of course; in the matter of tools the large and heavy are probably more durable than the light and flimsy. Perhaps the shopkeeper did not notice that Mr. B. has the pallor of the Londoner, and only stands five foot seven in his socks. The outfit of tools is designed for a giant with the muscles of a farm-labourer. An hour's digging or forking will knock Mr. B. up completely, and as for Mrs. B., she will not be able to handle those tools at all.

(*Verb. sap.*, the unaccustomed gardener will get on better with tools that are on the light side. Good iron-mongers keep different sizes, and even a size for women —very useful. Country ironmongers are much better than London stores for such purchases, though they may be a trifle dearer.)

A week elapses before the tools can be delivered from the town. There is no hurry about starting on the garden, for the ground is as hard as iron with drought. Mrs. B. has ample time to acquire what she considers a suitable gardening outfit.

None of those pretty linen aprons for Mrs. B., dainty little woman though she is. She means to be a real gardener. She buys women's slacks, and stoutish shoes, and those gloves with all the edges on the outside. She fancies herself in her slacks, and wears them for dead-heading the roses in the front of the house. Two village boys pass. They know there are strangers in the Lodge, and to them a "foreigner" from London is supposedly as deaf and stupid as a foreigner from the Continent.

"Look at that, Bill. Wot is it?" says one.

"Search me," says Bill. "Looks like nawthen so much as the backside of a nelephant."

Mrs. B. turns beet-red and hurries indoors. When she looks at herself *back view* in her long mirror, she finds that what Bill said is only too true. But she has paid heavily for the ungainly garments, and with all the expense of the move to the country she could not afford to lose them, give them away or otherwise abandon them, too large, too loose and too long though they are.

The B.s lose patience with the drought and try to begin on their patch of land before it breaks. Mr. B. jars his wrist and knee, jabbing the big four-tine fork into the cement-like ground, and cannot by any means lift or turn over his forkful of earth. Mrs. B. thinks she might do better with the hoe, but jars *her* wrist so badly that it troubles her all winter, and leaves the roots of the weeds firmly embedded in the ground anyway. It is annoying to have to wait, in such lovely weather, and do nothing, when there's a war on, and everyone ought to be doing something. The bodies and spirits of the B.s are keen to subdue that patch of ground to the national need.

At last October and the rain. And how! as I believe they say in America. After the first soaking shower, Mrs. B. joyfully dons her elephant's-backsides, her thick shoes and her gloves, and hurries out with the large spade, her spouse being occupied in following his profession up at the Hall.

The spade is nearly as large as Mrs. B. She prods it into the now yielding earth, stamping all her slight weight upon it. Bearing back, she sees with pride that a large spadeful is becoming detached from the rest of the world. But can she lift it? No.

She brings her not inconsiderable wits to bear on the problem. What would she do with the lump of earth if she did lift it? Put it down somewhere else? Pretty silly:

and it is full of weeds too. How would it be if she took the fork instead of the spade, and tried to do something about the weeds? It might not be the brightest thing to do, but it could not be wrong.

The fork, though heavy, is rather more manageable than the spade, because instead of fetching up great hunks of earth, it breaks the earth up as it comes. Mrs. B. begins methodically in the top left-hand corner of her garden, and is thrilled at the amount of weeds which she throws out with every lift of the fork. Great long strings of whitish roots, with no ends—these belong to the forest of rather low-growing green leaves which cover that top left-hand corner. They come out in wreaths and festoons and knots, and Mrs. B., delighted, jabs and stabs like the recruit who took too kindly to bayonet-practice.

She has nearly reached the end of her first line of work when the rain comes down. She desists for a moment. But what, she asks herself, is the use of a gardener who desists every time it rains? She looks down at her foes, the coils and bunches and provocative half-buried ends of roots, and feels that she could not stop work at this point for the Flood itself.

After twenty minutes, she has forgotten that it is raining. The water pours all over her, but she is only thinking of roots. Her gloves are two cakes of mud, her shoes two larger cakes, while her trousers are boards of mud to the knees. She will be sorry, but she is not sorry now. She is mad with battle, and only slightly puzzled as to how to avoid walking on the work she has done.

After half an hour, a voice assails her consciousness. "You should work backwards," it says.

She looks up through her streaming hair, to behold a weather-beaten person of indeterminate sex, the very dyed-in-the-wool countrywoman who is established in the imagination of every town dweller.

"I came to call," says this visitant, "but saw you round the corner of the house. You going to discipline this piece of ground?"

"We're going to grow food," says Mrs. B. confidently.

"Um. Bishop-weed, spear-grass, nettles, sheep-sorrel down the lower end. It will keep your fork busy for a while, before you can sow anything. Do you mind if I tell you a few things?"

"I ought to work backwards? Yes, I see that now," says Mrs. B.

"And with a smaller fork. You'd be more thorough. And, if you'll forgive me, in different clothes."

"Oh, what clothes?"

"What I wear myself," says the weather-beaten lady (who is only one of a dozen weather-beaten ladies of my acquaintance. We all get like that in time) "is a short coat of any description, if waterproof so much the better, and *men's* flannel slacks tucked into rubber boots. It's the only outfit. Perfect freedom of movement, a good deal of protection, and one keeps reasonably clean. Nothing is spoilt, even on the filthiest day. Nothing gets dirty but the boots, which can be washed at the pump or with the car-hose. As to you, you're a wreck already."

"What should one use for a hat?" says Mrs. B.

"A flying helmet: no brim to shoot water all over the place. I see you wear gloves—if you can call them gloves. I'm sorry about that. No real gardener wears gloves. You couldn't plant or prick out in gloves—how

could you feel if you had made nice comfortable holes?
I occasionally wear them, in arctic weather, but my
hands feel like feet in them."

"How do you keep your hands decent?" asks Mrs. B.
Her visitor's hands are innocent of varnish, but perfectly
clean, and indeed white.

"Oh, I just don't," says the countrywoman. "I never
set up to be an authority on how to keep the hands
lovely while holding down the job of gardener-of-all-
work. I am content to keep clean. That's easy enough.
Earth isn't dirty like oil or grease. Soap and water and
a stiff nail-brush are all you need."

"A stiff nail-brush," murmurs Mrs. B. in troubled
tones. Like most Londoners, she has been taught by
beauty-purveyors that a stiff nail-brush is to be avoided
like the plague.

Her visitor eyes her balefully. "A stiff nail-brush. Of
course, if you really mean to be a gardener, you won't
dream of varnishing your nails. Useless, and in bad
taste besides. It's a thing I have never done in my life,
I am glad to say. Barbaric."

Mrs. B. is glad that she is still wearing her mud-caked
gloves.

★   ★   ★   ★   ★

October is a month of frightfulness on the part of
the climate. Nothing daunts Mr. and Mrs. B. except,
at length, the suspicion that they may be doing more
harm than good by tramping about on their sodden
piece of ground. By the third week of the month the
place is clear of weeds, to outward appearance. Yards
and yards of roots of bishop-weed, otherwise ground-
elder, have been dug out, and hundreds of yards of the

threadlike roots of the small sorrel from the other end of the patch, not to mention great fibres of nettle and long fangs of thistle, and mountains of the soft green of winter-weed. This one piece of ground seems to be afflicted with all the ills that gardens are heir to. But at last it looks clean, to the B.s.

What next? It must, of course, be dug over, from start to finish, with a spade, to undo what the tramplings have done. Mr. B. invests in a smaller spade. It takes a smaller bite of ground, and so is slower, but he can work longer with it. The larger one incapacitated him utterly with blistered hands after two hours.

He studies how to dig. Not any fool can dig. Mr. B. is no fool, and has some idea of shoving the spade as straight down as possible, so as to make it do its maximum of work at each effort. But he asks himself just what his wife asked herself some time ago. What does one do with the spadefuls of earth?

Why, in the first place, does one dig? To loosen the soil, to aerate it, to—ah, thinks Mr. B., I knew there was something that went with digging like sausages with mash: digging and manuring, old chap, digging and manuring. First catch your manure, and then make a place to put it, by digging.

Casting down his spade, he gave himself up to the problem of manure.

And here the weather-beaten lady, being a native, was able to give advice. "Farmyard manure is best. It has everything you want in the way of fertiliser, and something besides—the straw in it is very good for lightening heavy soil. The trouble is that farmers need manure themselves. They don't want to be bothered with selling and delivering one cartload to some potterer with a

5 FORKING UP THE BOTTOM OF A TRENCH, FOR GOOD DRAINAGE

6 HOW TO DIG: BLADE OF SPADE STRAIGHT DOWN

7 SOWING (a) RIGHT AND (b) WRONG

8 RAKING THE SOIL TO A FINE TILTH

garden. Since you've got to have manure here and now
—the autumn's getting on—you will have to go on your
knees to someone, offering a bag of gold at the same
time. But before next year, you shall have some of
your own."

"Eh?" said Mr. B., with his mouth open.

"It's easy. If you look in the gardening columns of
any reputable paper such as *The Times,* you'll see now
and then an advertisement of one or other of the sub-
stances which hasten the breaking-down of garden
rubbish into manure. It doesn't matter which you decide
to use; you can even do quite well, but not so speedily,
without any. Merely dig a large hole in a corner of your
garden and put into it leaves, grass-cuttings, potato
peelings, outside cabbage leaves, seeded lettuces, anything you've got bar weeds. Not too large a proportion
of autumn leaves, because manure ought to be richer
than leaf-mould. You'll want another pit for leaf-mould.
You're frightfully lucky to be surrounded by trees."

"Frightfully," agreed Mr. B. with enthusiasm, and
then reflected that someone would have to sweep
up all the leaves before they could become manure or
mould.

"The best way of all," said the weather-beaten lady,
"is to keep a pig."

"Oh, I don't think we could keep a pig," said Mr. B.
Gardening wasn't farming, after all.

"You may change your mind when rationing comes
in," said the weather-beaten lady, and Mr. B. experienced a shock. He had forgotten all about the war.
His mind was on food-production, and not on the
reason for it.

She told him of a farmer who was likely to be less

hard-hearted than most in the matter of delivering a small quantity of farmyard manure. "Ask for horse if possible," she said. When the load arrived, it did not look a very small quantity to Mr. B. He was a little shy of it. He did not like to think of its origins; he did not care for its smell.

But very soon, in the interest of digging it in, he had overlooked these drawbacks. He had evolved a plan for getting the soil out of the way and the manure underneath it. He dug a trench across one end of the plot of ground, and carried the soil away in a barrow to the other end. He felt slightly childish doing this: there must be a better way, he was sure. But it *was* a way. Into the trench he forked a layer of manure: then stepped back a pace and dug another trench, covering up the manure as he did so. In this way, he dug and manured the whole piece; the soil that he had carted out of his first trench was needed to fill in his last.

Later, he saw other diggers working in other ways. One way was to spread a piece of ground with manure and simply turn the spadefuls over, manure and all. It needed practice, and considerable strength. Mr. B. remained convinced that his own way was more thorough. Another thing he saw which shocked him: the gardener-boy up at the Hall would turn over the spit of soil *weeds and all*. He did not fork the weeds out first. Mr. B. took this matter to his weather-beaten friend, and she agreed with him that the practice was bad. "It's done in the fields—the plough just turns everything in—but a garden isn't a field. There's less room for weeds in a garden. Another thing: many of our garden weeds are perennials, and many shoot from running roots. You might as well sing as hope to eradicate them just by turning them over. It will work for annuals, but not for perennials."

Mr. B. asked his mentor if she thought his plot of ground was now ready for sowing or planting: and if so, with what? She said the weather was too wet; one needed moisture, but not a sodden earth. She suggested that the B.s should wait for at least twelve hours' dry weather before setting any plants or seeds. If there was no dry day before November, they would have to put a few things into the morass to take their chance. "But meanwhile," said the good lady, "there's a thing you could do. The lower end of your plot was riddled with sheep-sorrel: that means sour light land. You could find some marl or chalk, and spread it over the top. No need to be very accurate as to quantity: scatter it so that the soil shows through. There are chalk-pits and marl-pits not far away: luckily, in most parts of the country, you find chalk-pits and marl-pits here and there."

Soils were a closed book to Mr. B., as they are to most people who do not have to dig in them. To the hand that works a garden, there are enormous differences in the textures of soils. One deems oneself fortunate in many respects, though not in all, to have a light soil, or sandy loam; this makes for deliciously easy working, and one can carry on with work in most weather except a severe drought, whereas on stiff clayey soil one does more harm than good by stamping it down in the wet. The light soil, however, needs more feeding; two barrow-loads of farmyard muck to the square rod is not too much. The muck should be well-rotted and not too strawy, for light soil is well enough aerated without this; furthermore, leaf-mould is not nearly so useful on light soil as on heavy. With very light land, the problem is how to conserve moisture; it is sometimes necessary to dig-in clay, but this is a difficult job, and as a rule the farmyard manure will suffice.

Very light soil responds well to digging or trenching in spring, but there are of course certain vegetable crops which we want to start in autumn, so some of the garden at all events must be prepared then. Heavier land should all be done in autumn, and the top left rough after digging (except where seeds are to be set) for the winds and frosts of winter to deal with. If winter does its duty, there will be a pleasant fine crumb on the surface by the time we want to do our spring sowings.

There are many artificial manures available for those who find farmyard manure difficult to obtain. But artificials do not quite do the work of the real thing. They fertilise, but they do not fulfil those other two important functions, aeration and moisture-conservation, nor do they restore the humus. I have never used artificials in my garden, although, as I am merely a wife and don't offer to pay for it, I have the utmost difficulty in securing a proper amount of manure from my husband's farms. It is a great help to have a domestic animal, such as a pony or pig, on the premises. The revival of pony-driving, owing to petrol-restriction, will serve two purposes if the pony is taken in from the paddock for the night or part of the day, and kept in a stable and adjured to do his duty by the garden. Pig manure is rather strong, and must be allowed to rot thoroughly in the muck-heap before use; in fact no natural manures should be used too fresh.

People who are beginning gardens in a hurry under the stimulus of war may be positively unable to obtain

farm manure, so here are some hints on "artificials."
Kainit is a cheap source of potash (though it may become
scarce as it is imported largely from Germany) but is
only suitable for light land as it causes "panning" on
heavy land. It should be applied 3 weeks before sowing
seed. For heavy soils use sulphate of potash. If basic slag
is used as a source of phosphate, it should be applied in
autumn, as it is slow in action. The quantities of these
are: Kainit, about 5 lb. per rod, sulphate of potash 3 lb.,
basic slag 7 lb. But a well-balanced manure may be
mixed, with a little trouble but better results, as follows:

> 1 part sulphate of potash (or 3 parts Kainit).
> 1 part sulphate of ammonia (or 1 part nitrate of
> soda).
> 3 parts superphosphate.

This may be applied at the rate of 7 lb. per square
rod. If some farmyard manure, but not enough, is
obtainable, use up to half the quantity of the balanced
artificial mixture as well. It is not advisable to apply
farmyard manure to ground intended for carrots or
parsnips in the same year that they are grown, but if
they are being grown for the first time in a new or
neglected garden, the balanced artificial ration will suit
them well. A fortnight or three weeks should elapse
between manuring with artificials and sowing.

I see that we have now arrived at the stage of men-
tioning vegetables by name. At this stage also have Mr.
and Mrs. B. arrived. Their plot is clean and dug, mucked
and sprinkled with marl—and the rain it raineth every
day. While they sit by the fire and look out of the
streaming grey windows, they say to one another out
of periods of profound silence, "Artichokes. Do we like

artichokes?" "Beetroot: how much food value do you suppose there is in beetroot?" "If you want vitamins, carrots are just bursting with them: the monthly-nurse I had for Jessica was crazy about carrot-juice." "Yes, I remember," says Mr. B., "and didn't poor Jessica have wind! What I'm really wondering is how to induce the children to eat our precious vegetables. The Government has offered some hints to those housewives who are catering for evacuee children, and it practically admits that no child likes vegetables."

"Don't tell me," says Mrs. B. severely, "that you are *going off* gardening?"

DIGGING FORK

II

WHAT TO GROW: AND WHEN, HOW AND WHY

"Why not potatoes?" says Mrs. B. to her gardening friend. "Surely potatoes! We eat more of them than of anything else."

The weather-beaten lady (who is regrettably didactic, I admit, but who saves me from having to be didactic in my own person all the time) replies, "Potatoes are a farm crop, and are being looked after by Government. I'd be surprised if there was a shortage. In normal times, every potato we need is home-grown, and only the luxury earlies are imported: and if there are to be more potatoes used now as substitutes for other things, the Government will know how to stimulate farmers to grow them. Besides, my dear, it's futile to grow potatoes unless your cultivated land is more than an acre in extent. I'd as soon grow wheat in a small garden as potatoes: it would contribute just about as much to the national economy."

Much crushed, Mrs. B. awaits further pronouncements.

"I should say," my mouthpiece continues, "that the shortage will be felt in other vegetables, just because of the Government stimulus to main crops. Farmers who have grown occasional fields of cabbages, sprouts and peas won't do so any more. There is, too, the distribution problem—market-gardeners may not find it easy to contact all their usual retailers. You are going to need more vegetables, to make up for a possible shortage of meat and fish, and you are not going to get them through the usual channels. You must grow them."

That is what the B.s mean most earnestly to do. The defeat on the potato front is a slight disappointment, but there are plenty of other vegetables.

"Somehow," broods Mrs. B., "one never looks on other vegetables as *food*."

"Usually," says her friend, "they aren't food. They're fibres and water, ruined by bad cooking. But that is another story."

We are now called upon to exercise imagination. We are to imagine that the weather holds up sufficiently for Mr. and Mrs. B. to do some autumn planting. By dint of consultation with everybody they see, they have drawn up a sort of rough calendar for their garden. There is nothing more foolish than a hard-and-fast calendar, but the B.s are quite right in trying to make an approximate time-table for their activities. There are so many different things to do in a garden that it is quite easy for the beginner to miss out an important crop altogether.

Most vegetables begin their careers in spring; but if you want cabbages in late May, when there is a great scarcity, then cabbages should be planted out before the end of October, or very early in November, should October be impossible. The inexperienced gardener, and the town-dweller generally, hardly realises that the greatest shortage of vegetables occurs in late May. It seems that everything should be plentiful then, in the abundance of spring: and for those who just pop round to the shops, everything is fairly plentiful at all times, since what is not grown at home is imported. But the country gardener knows when the shortage occurs: and if, owing to the war, imports of French vegetables are much reduced, townspeople will know it too.

Our friends the B.s, having no cabbage seedlings of their own, have to buy some from a local market-gardener. Next year, they promise themselves, everything will be home-produced from the seed upwards. They plant out the seedlings 18 inches apart, following the instructions of their knowledgeable friend, and leave the rest to God. After which, they set broad beans, longpod variety, for an early crop, pushing the seeds into the ground about 6 inches apart. It is hoped, when one sets seeds separately at this distance, that every plant is going to grow; there is no thinning-out of broad beans. The distance between rows should be 2 feet; but as a matter of fact, Mr. and Mrs. B. only set one row of earlies, because they hope to avoid monotony in their diet. In January they will set some more.

They then cast round for something else to do, and think of rhubarb: surely that is a plant which stays in the earth all the year round. Nobody likes it very much, except at its youngest and most delicate, in spring: but if it is going to be wanted in spring, something must be done about it in November. (We must bear in mind that the plot of garden at the Lodge is totally empty.) Mr. B. has seen a forest of gigantic rhubarb in a neglected corner of the garden at the Hall; by making friends with the elderly one of the skeleton staff up there, who is always willing to talk of the past glories of the place, he is able to "borrow" two crowns, with good fat buds on them. The old man suggests an extra deep trenching for the ground into which they are to go. He would have liked to part with more than two crowns, but Mr. B. thinks of the large amount of space taken up by the plant, and declines with thanks.

In the vast kitchen garden at the Hall are fruit trees. Mr. B. would dearly like to have fruit in his small plot: apples, at all events—so good for the children. But the room taken up by the standard trees at the Hall seems to rule them out as denizens of the Lodge. He broaches the subject with the old man.

"You could allus hev cordons or 'spaliers," says that sage, "if you think this here war's gooin' on that long."

"Perhaps," says Mr. B., "even if the war doesn't last the expected three years, we shall have acquired a taste for country life. I think we've acquired it already."

"What? You hain't had nawthen only rain the best part of your time here. Even us country bu'ds is fed up."

"A spot of rain is nothing," declares Mr. B. (We are supposing, it will be remembered, that the weather is now clearing.) "After all, we aren't holiday-makers, we're gardeners. Tell me about cordons."

He learns that apples and pears take up a minimum of space in a garden, and bear fruit heavily, if trained on walls and fences. "And they look that pretty," says the gardener. Pretty? Mr. and Mrs. B., in a fervour of utilitarianism, had forgotten that the first duty of a garden is to feed the soul. "Whoy don't you," says the gardener, "make a path down the middle of your piece of ground and set low-growing 'spaliers beside the path? At the edge of the walk you could put in parsley and mint and sage and all them things what gets forgotten and is so badly missed."

Mr. B.'s garden ceases to be a thing of straight rows. It takes on grace in his mind. It will have a girdle of blossom.

"You want to git the trees as soon as you can," warns the gardener. "November's a good month for anything woody. But if they fare to need a lot of twisting, like, to git them to lay along their wires, let the twisting wait till spring-time, when the limbs is tender with sap."

Mr. B. leaves the gardener, his mind full of stakes, wire and apple trees. He is recalled. "Pardon me, sir, for hollerin', but I'm too stiff to run arter yer. Don't you muck them trees, not with yard muck, or you'll git leaves 'stead of apples. Dig you deep, to drain the land, and put in leaf-mowld or what you like, but don't plant 'em right on to muck. Muck 'em round in bearing-time, that's what suits 'em."

"Oh, and what sorts should I grow?" Mr. B. suddenly remembers to ask, as he is turning away again. Gardening does tend to prolong conversations.

"Don't grow earlies, that's my advice. Everybody grows earlies, and if you want to eat a few, you can allust steal 'em. Grow keepers, what you can eat when there ain't nawthen else. Cox's, or Ellison's Orange, which is just as good, and a bit bigger and earlier. That supplies you for November. Laxton's Superb for Christmas and after. For cookers, there's one called King Edward VII which do well on a cordon and keep nicely. Allust grow more'n one sort, cos some on 'em don't fertilise theirselves."

As Mr. B. at last withdrew out of earshot, he was followed by the warning, "Don't expect to see no return on your money next year, nor not a lot the year arter."

Perhaps, then, apples were not such patriotic things to grow: the war would be over before they began to produce food. Mr. B. set his jaw. He meant to have a jolly good shot at going on being a countryman after the war. He tensed his newly-sprung and magnificent biceps, and spread his broadening shoulders. He thought of the grand rhythm of digging: strike, shove, *heave*, strike, shove, *heave*. He thought of the chocolate earth, that was so satisfyingly dirty in such an essentially clean way. He made plans for after the war. Even supposing that he had not made or inherited enough money to retire upon, he could leave his wife and family here at the Lodge and spend every week-end digging. Why *not* have apples up the middle of his garden? It was as good a speculation as any, and better than most.

The apples took him the rest of November to deal with. Seeing that the autumn is the time of preparation and the winter the time of quiescence, one must not expect to be doing a great deal of original, creative work (so to speak) in November. It may seem that I have begun my account of a gardener's year at the wrong end: but I think not. The preparation is the real beginning of a garden: not the sowing. The sowing cannot be done unless the preparation has been done first. The farmer's year begins in October, and so does the gardener's: and it so happens that by the caprice of Herr Hitler, a very great many people have been enabled to begin food-producing at the proper time. In an open winter—the ideal winter is one in which spells of frost alternate with mild showers and periods of sun—the operations of digging, weed-forking, and mucking can be carried on in December and January; I mention this for the comfort of those who started late and became

9 ESPALIER PEAR

10 CORDON APPLE

11 YOUNG APPLE WITH PRUNING-PLACES MARKED BY TIES

12 SUPPORTING YOUNG APPLE: NOTE THICKNESS OF BINDING

bogged in October. But beware of digging-in half-frozen clods. Frost is a wonderful weed-killing, pest-killing and pulverising agent, and should be given a free hand.

Pruning of fruit trees may be done any time from late October to February, and there are decided advantages in doing it early; one can more easily tell what is really dead and diseased wood before the whole tree takes on its winter look of deadness. All prunings should be burnt. Good gardeners make a rule to burn everything that is not convertible into manure. Woody prunings are not; moreover, there is no knowing how long a diseased shoot may go on carrying its disease in active form. Clear away prunings, then, and do not put them in the compost-pit. Neither spare from the bonfire any noxious weeds—they are incredibly tenacious of life, and will leap up again as good as new if applied to the land in the form of compost.

Pruning, says Mr. Middleton, is largely a matter of common sense. That goes for most gardening operations. With apples—since apples are now exercising the mind of our friend Mr. B.—it is necessary to clear the middle of a bush or standard tree of all small growth: apples are produced on the outer shoots, not in the middle. Of this year's outer shoots, if they are very plentiful, shorten some and leave others. Young trees not at the age of bearing should, of course, have all their shoots cut back half-way; it is even advisable to prune a tree in the very autumn of planting. As for Mr. B.'s cordons and espaliers, in the case of cordons all side shoots should be reduced to about an inch long— the tree bears apples all down its single long stem, close to the wood, on short side shoots, and to cut them back makes them more numerous. Also take a few inches off the top shoot. For espaliers, the pruning is directed to making the apples form on short shoots along the lateral branches. But young trees must not be allowed to fruit; resisting all temptation, we must remove any flower-buds in spring; three years old is early enough for an apple tree to come into bearing.

Use a sharp tool—knife or secateurs—for pruning, and cut close above a bud.

Most young fruit trees are the better for a stout stake planted close to the stem and secured to it with strips of rag 2 inches wide, or with anything that will not fret the bark. Secure in several places, for a high wind may break a single tie. Our Mr. B., of course, has a whole fence to support his espaliers.

If he had cared to, he could have grown pears as well on the cordon or espalier system, but he knows his limitations, and thinks chiefly of dietetic usefulness. Pears, however they are grown, need the same sort of pruning as apples. Plums, on the other hand, need very little: merely, when young, they must be pruned with an eye to shape.

Here we are, then, at the end of November, and Mr. and Mrs. B. are all set for food-production in the spring. There is little more that they can do at the moment, so they busy themselves in collecting information and making additions to their calendar. They wish they had begun gardening *last* year. "We'd be eating our own celery, if we had any celery," murmurs Mr. B.

"Next year we'll have celery; it's good for rheumatism," says his wife.

"But we haven't got rheumatism."

"We shall have, if next autumn is like this one."

*   *   *   *   *

We might find it helpful, in the winter quiet, to look over the shoulder of Mr. B. as he makes his calendar. These serious amateurs, they do take pains.

"Sweep up leaves" is an item that occurs in October, November and December. This activity is determined, naturally, by the season. In some years the trees hold most of their leaves till December. It is foolish to sweep up a few leaves at a time; on the other hand, it is inadvisable to let leaves lie in thick drifts on a garden bed for more than a week, if there should be anything planted in the bed. They are too warm a covering.

Leaves, being swept up, go to one of three destinations, the manure pit, the leaf-mould pit, or the hotbed. This last is an ambitious move on the part of Mr. B. He means to bring on early salads in a frame. To this end he is making a neatly squared heap in one corner of the garden, composed of leaves, some yard manure, and more leaves, which will eventually be three feet high, before it is topped with about twelve inches of earth. In the making stage, it is turned over with a fork every two or three days. The village carpenter will make a framework to fit the top, with a sliding lid of panes of glass like a sash-window.

Another December note is, "Mustard and cress on flannel in the nursery." The B.s have no greenhouse. They are negotiating for a lean-to; they will not be able to heat it, fuel difficulties being what they are, but if it leans on a house-wall and faces south, it will be better than nothing. For the present, however, "Mustard and cress on flannel in the nursery": a way of combining amusement and instruction with vitamins.

"If we had our lean-to, we'd sow onions in December,"

sighs Mrs. B., bored with the enforced slackness of December.

January if fine and without frost is a promising heading. More broad beans may go into the ground. People with greenhouses may sow lettuce. Artichokes, the root kind, may be planted outdoors, but only in a large garden, for they make huge plants. Rhubarb should be covered up (whether January is fine and without frost or not) to produce early and tender stems: straw should be loosely bunched over it, and the whole covered by a bucket without a bottom, such as is to be found in most garden sheds; or a box with the bottom out, or if you are very highfalutin, a proper forcing-pot bought with money.

Brussels sprouts may be sown under glass by those who have glass. Likewise cabbage. Sow thinly in crumbly soil; if your garden soil is stiff and claggy, mix in some sand and leaf-mould. The best people put their potting-soil through a garden-sieve; I work it and crumble it with my hands, merely because I like it.

In a dry spell, any time now, woolly aphis may be dealt with on apple trees, It looks like thin tufts of cotton-wool on the bark; and underneath the white fluff are the creatures. A stiff brush, which really gets into the crinkles of bark, used with methylated spirit, will control all but the worst cases.

February.—All gardeners hope that spring is going to begin for them in February. In a favourable season it may be a very busy time. Before the month is out, some more broad beans may be put in (we hear a good deal of broad beans because they are truly hardy. Not everyone will want to plant broad beans every time I say so: personally I consider them delectable if eaten

15 SIMPLE FRAME ON HOT BED

13 SOWING PEAS

14 PRICKING OUT YOUNG LETTUCE IN A BOX

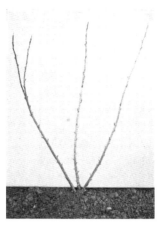

16 YOUNG BLACK CURRANT
UNPRUNED

17 PRUNING BLACK CURRANT

18 YOUNG BUSH APPLE UNPRUNED

19 PRUNING YOUNG APPLE

very young and properly cooked). As well as the re-
current bean, peas may be planted in the south—not
less than 2 inches deep. Turnips may be sown, and
parsnips are all the better for an early sowing. I have seen
a rather special method used for sowing parsnip: having
dug deeply, make a deep hole with a pointed stake,
pushing it round to enlarge the mouth of the hole; fill
this with finely crumbled earth and drop two or three
seeds in, covering lightly with soil. When the time for
thinning comes, leave only one plant in each hole, and
you will have a long and shapely root. Plants should
be about 6 inches apart, with 15 inches between the rows.

Outdoor onions may have their bed prepared: they
like a fine tilth. Spinach may be sown. This is another
of those recurring decimals; one needs successions of
sowings, if the size of the garden permits. Peas, too,
should be sown as often as may be: every three weeks
if possible. On a real gardening calendar, not a con-
densed one such as this, these recurring duties are written
down every time and marked off when done; this
ensures that they are done often enough, though not,
perhaps, on the day appointed. Man proposes but the
deep depression from Iceland is the real master of the
situation.

March.—Now we get busy. Anything may be sown.
Our onion-bed has already been finely raked; the seed
must not be buried more than an inch deep, and if it is
sown in rows, the thinnings can be used as spring onions.
Some people roll or otherwise make firm their onion-
ground after sowing, but it is not advisable to do so
to any but light land.

People who, unlike Mr. and Mrs. B., are well established in the gardening world, will have onion-thinnings

for their salads already, from the rows which they sowed in early autumn. The onions should be left at least 6 inches apart.

Country people always plant their potatoes on Good Friday, whether that day falls in March or whensoever —unless there is actually snow. I can't say much for this practice, which savours of superstition rather than sense, but it does prove that in a mild year, potatoes may go in in March. Mr. and Mrs. B. are not growing potatoes, nor would I advise any owners of small plots to waste space on something of which there will probably be no shortage. But owners of large gardens take pride in being able to proffer a dish from their own gardens, and there is no reason why they should not. They should plant only early kinds in March.

Carrots may be sown now: be careful that they are not on land where they were last year, and that their land has not been recently manured. They like a light, sandy soil worked to a fine tilth, and when the time comes they must be thinned very thoroughly, to at least 6 inches apart. Some of the thinnings will be big enough to use, raw, in salads. Lettuces and radishes may be sown; they should be kept going all summer, if liked, by fortnightly successions of sowings; radishes mature quickly, in damp weather, but are not nice if they happen to strike a dry time for growing. Eat them young. They take up very little room in a garden and do not occupy their place for long. Mustard and cress—to continue the salad *motif*—are difficult to raise satisfactorily out of doors, as they nearly always grow too tough and sturdy. The damp flannel on saucers in the nursery is the next best thing to a box in the cool greenhouse.

Parsley may be sown about now. But it will only

grow in gardens where the missus is master, so they say! Although for some people it will flourish exceedingly, I have never managed to produce any. And it is such a useful thing to have, not only to make a mutilated joint of cold mutton look like lamb, but to improve the taste of potatoes, if chopped fine, mixed in melted butter and poured over, and to make a white sauce less like bill-stickers' paste.

Go on sowing peas in three-weekly successions.

April.—In large gardens, asparagus crowns may be planted. Asparagus seems to have been a sea-plant: it likes a sandy soil and a top-dressing of salt. It entails a good deal of work, and not a little expense in the original lay-out, for unless one means to wait a couple of years with one's mouth watering, three-year-old crowns must be bought. An asparagus bed is difficult to keep free from weeds, as the young shoots come up at the very time when one would like to be hoeing, so here again is expense either of time or money—some-one must do the weeding by hand. In order to make the surface of the bed more accessible, it is advisable to make it into a hump, like a long grave. Plenty of growers grow asparagus on the flat, in fact when it is produced in fields one seldom sees humps, but I have always seen good results from humps, possibly because there is more of the bed-surface exposed to the sun. The shoots may be cut when they are 3 inches out of the ground; with established plants, this may be in mid-April in a warm spring. The best tool to use is a special knife with a saw-edge, which cuts an inch or more below the soil and does not cut the top off the adjacent baby shoot. (Cutting should cease at midsummer, for the plant's sake.)

Sow now cabbage, cauliflower, broccoli, sprouts,

savoys and leeks, to be transplanted when large enough
to handle; also turnips, spinach and lettuce. Forget not
the successional sowings of such things as we have
already mentioned. Celery may be sown in boxes in
the cool greenhouse. Mr. and Mrs. B. must have their
lean-to by April, or it will not be much use to them this
year.

As April goes forward, some of the box-sown seeds
will be found to have grown enough for planting-out,
and the earlier outdoor sowings such as carrots will need
thinning.

But the thing which most exercises us in April is
weeding. All those weeds which Mr. and Mrs. B.
thought they had removed from their garden-plot and
burnt are resurrected on the first warm, damp day; and
thousands more, the seeds which dropped on the soil
unseen during the whole of the late summer and autumn.
April is a shock to Mr. and Mrs. B.—they can hardly
see their vegetable-rows for weeds. They think of the
hours they spent with the big fork, and the blisters it
gave them, and wonder if it was worth doing. Assuredly
it was. Our spring weeds are for the most part annuals.
The wicked perennials are certainly scotched, though
not killed. The swan-necked hoe will deal with weeds
in spring.

May.—Two excellent vegetables are sown this month,
the marrow and the runner bean. Marrows do not
tolerate frost, so be guided by the natives as to the
weather prospects. The ground where an old manure-
heap used to be is a favourite spot of theirs; they grow
quickly and rankly, and cannot be allowed to sprawl
among orderly rows. They produce far more marrows,
as a rule, than can be eaten as vegetables, but much later

20 PLANTING OUT ONIONS

21 TYING COS LETTUCE

22 THINNING ONIONS

23 SOWING IN STRAIGHT ROWS BY MEANS OF STAKE AND STRING

24 SOWING SEED IN SHALLOW DRILLS

in the year, in autumn and early winter, when the juice has dried out of them, they make jam and chutney. To make the jam, use three-quarters of a pound of sugar to a pound of marrow cut into cubes, flavour with ginger or lemon as preferred, boil until transparent and inclined to set when tested on a cold plate. For chutney, throw in anything you have in the way of sultanas, peppercorns, orange-pulp—oh, well, anyone can invent a chutney and cook it with demerara sugar and vinegar, and this isn't a cookery-book, anyway. But just bear in mind that the marrow fills many a long-felt want. If it is thought somewhat insipid as a vegetable, try it *au gratin*.

And ah, the runner bean! It is to be hoped that a place has been reserved for it where its beauty may be appreciated, as well as its food-value. The fence at the end of the garden, or down the side—it looks well as a background for other things. But as it may grow to 9 feet in a damp season, it should be placed so as not to shut out the sun from everything else in the garden. It is a vegetable which I would recommend to everybody. Its food-content is good, and it supplies iodine. One sowing gives months of beans, for they mature in stages, those nearest the ground ripening first. As with all pod vegetables and a good many others, they should be used very young if they are to be at their most delicious. Those who employ gardeners seldom have young enough vegetables brought into the house: the gardener likes to see size in his product.

Runner beans need firm staking with large stakes. There are ways, in the country, of procuring faggots of small tree-branches from those who clear hedges and copses. In the suburbs, this is more of a problem: but if poles are procurable, a sort of triangular framework could be made with wire. In any case, the stakes should be set about 2 feet apart at the base, and leaning towards each other to touch at the top. Most kinds of peas are the better for staking also.

Beetroot may be sown in either April or May. There are ways of dealing with this vegetable other than the usual slices-in-vinegar method: it may form the basis of good winter salads. A row of it should certainly be put in.

Most of our vegetables, then, are sown. We can keep on sowing those that are marked as "successions" right on until we start sowing in autumn for spring. But for purposes of instruction, we have enough. A second sowing of runner beans, if it is needed, may be made in June, and the humble radish may be sown any time in damp weather. But except for these, we are all set. It is to be hoped that we have been wise enough to do our sowings when the ground has been neither parched with drought nor pasty with wet.

I have particularly avoided mention of out-of-the-way plants, the pride of the epicure. Life is going to be quite full enough for the beginner, without egg-plants, sweet corn, globe artichokes, celeriac and Scorzonera. Incidentally, I know nothing about them, except the delicious but space-wasting globe artichoke, which is very easy to cultivate, but takes up far too much room for its negligible food value.

Later in May, planting-out may be done of any cabbages, cauliflowers, celery or sprouts which have been raised under glass. When planting-out, make a hole large enough to hold the roots without compression, water the hole a little unless the weather is showery, set the plant in upright, and make the soil firm round it. That the plant should feel comfortable and firm is essential. One cannot ensure this by using a trowel: the hand is the only tool.

Meanwhile, in other parts of the garden, this is the time of grass-cutting. Remember the manure pit!

It is also the time, in certain seasons, for the arrival of greenfly. This pest does not confine its attention to roses: both green and black fly are fond of beans, and our broad beans must be looked to. I always keep a bottle of Paterson's Clensel at hand; it makes an inexpensive wash (instructions on the bottle), and what is not needed for the garden will wash clothes *à merveille*. Always use a non-poisonous wash for vegetables. Look out for onion fly on spring-sown onions: a dusting of soot is a good preventive.

A vegetable which I ought perhaps to have mentioned is the tomato: but in most parts of the country it is hardly worth growing out of doors. A note of it will be found in the alphabetical list of vegetables.

June.—Keep hoeing. Keep hoeing all summer. In places where the hoe will not go without damage, a small hand fork will do the work. The function is twofold, to destroy weeds and to break up the surface soil in order that it may hold moisture.

Plant out any seedlings that were not large enough to be handled last month. If potatoes are being grown, draw the earth over them in ridges, leaving a little green showing at the top. As the green grows, more earthing can be done, but see that the hump is broad and thick, not tall and thin; i.e. that you are not exposing any potatoes in scraping up the earth.

Plant out celery. There are two schools of celery-planters, the trench school and the flat school. The trench is more commonly recommended, so here are instructions: dig to a depth of 15 inches or more, and a width of a foot, lifting out the soil and making banks on either side. Remove still more earth from the bottom, and replace this by manure mixed with soil and, if possible, peat. (Celery likes acidity, so avoid lime or other alkalis.) Put some more good soil over the manure-mixture. The celery plants should be set 9 inches apart, and once planted should never be allowed to suffer from drought. When the plants are a foot high, earthing-up can begin—say in August, and every so often till October. The plant must either be held together with the hand, or lightly tied at the top, to keep the earth from between the sticks.

Leeks too may be put out now. They are treated like celery by some people, but it seems easier, presupposing that the garden has been deeply dug in the first place, to use the dibber method: make holes with the dibber, of a size that will hide the young leek except for an inch of its top. Drop the leek into the hole, drop a little earth beside it and make firm with a stick, but do not fill up the hole till autumn.

Very early potatoes may be ready to be lifted in June, and when they are all taken up, it is not too late to use their ground for white turnips or greens.

Thin out beetroot, taking great care that the roots of those that remain are not broken when you pull the thinnings away. In all dealings with beetroot, it must be remembered that they lose colour if the skin is broken.

Keep hoeing. Look at your fruit trees and bushes, and if they seem to have set for a big crop, put a mulch of

25 SAVOYS IN SUNSHINE

26 EARTHING UP CELERY

27 STAKING PEAS

28 STAKED RUNNER BEANS

manure round the base of the stems, or else keep them watered.

July.—Make rhubarb jam now, if you did not do it last month, and take off the flowering tops of the plants, to conserve their energy.

Sow turnips for early winter use, also spring cabbage.

Summer-prune fruit trees by removing all thin non-bearing shoots, and some bearing ones if the crop is heavy; bear in mind that your object is to let light and air in to reach the fruit. See if any of your part-grown crops need thinning. Sow onions for spring. You can still go on sowing lettuce, stump-rooted carrots, etc., and if there is some ground cleared of leguminous crops (beans or peas) so much the better. I am reminded to mention that leguminous crops not only feed the inner man, but feed the ground as well, by means of the nitrogen which they store in those warty-looking nodules on their roots. As far as the inner man is concerned, these plants give him protein, and so supplement the meat-ration. Those who remember the last war on the home front will remember cutlets made of bean-flour, of split peas . . . everything was called a cutlet, and all cutlets were made of beans. And we survived it.

August.—Some sorts of potatoes will be ready to take up. Choose a dry day, and let them lie in the sun for a time, so that the earth will come off easily. Store in a "clamp": that is, make a long heap of potatoes—if the ground is ill-drained, let the bottom of the heap be a platform raised three inches or so from the surrounding land—and cover them with clean wheat straw, then cover the straw with six inches of earth, well patted down, and make ventilation-holes in the top of the ridge, putting a loose twist of straw through each to keep the frost out. Do not clamp potatoes too early; the end of the month, or later, will do.

Onions may be lifted when large enough, and dried in the sun, then hung on a wall in bunches, in a place where they have protection and air. Cauliflowers ought to be large enough to eat. Celery will want earthing-up. At the end of the month, cabbages may be sown out-doors. Vegetables from the later succession-sowings will need to be thinned or planted out, as the case may be.

September.—Sit back and enjoy life, more or less. Except in a wet summer, the weeds will be giving you a rest. But the wasps are out, and they will damage pears and apples, hard-skinned though the apple is. Look for nests and destroy them by poking pieces of cyanide down the entrance holes and then pouring in water, or laying rags soaked in paraffin in the entrance holes and setting a light to them (do not, in this case, completely stuff up the holes, as the fire needs a draught. And in either case, run for your life). For local measures, hang jampots on the fruit trees, with just enough beer in the bottom to drown the brutes.

Young marrows are ready to be eaten.

Go on lifting potatoes and onions of the later sorts. See that celery is nicely earthed-up. Leeks can be earthed-in at the end of the month.

And late in September you can begin all over again with the digging.

And here we are back again at *October*, where Mr. and Mrs. B. began. But this year we are not dealing with a garden empty of all but weeds. There is beetroot to be lifted, with carrots and parsnips. The latter should be left in the ground till a frost has touched them, but beetroot will not stand frost. All root vegetables keep fairly well in boxes of sand, or simply in heaps of sand deep enough to cover them, in a frost-proof shed. Turnips, however, taste better if eaten straight from the ground, and need not be stored.

Apples (supposing that your trees are not as young as those of Mr. and Mrs. B.) are gathered and stored in October. Storage should be in slatted wooden trays in racks, and in the dark. The fruit should lie in neat rows on the trays, not touching, and all fruit that is not perfect should be eaten at once instead of stored. In some years, this means eating apples till one is sick of the sight of them; as a change from the everlasting apple-tart- or stewed-apple repertoire of the everlasting cook-general, may I suggest apple fool (made with Bird's custard if war puts whipped cream out of the menu), blackberry-and-apple "cheese" made by stewing the two together with sugar, putting the result through a sieve, and adding gelatine in the proportion suggested on the packet; and the addition of one sweet apple and one sour to the winter salad that can be made with beetroot and celery cut into dice (and potato and onion and whatever else one fancies). There is also the homely bread-and-butter pudding made with layers of sliced apple and brown sugar alternating with the layers of bread and butter; this is just as nice made with margarine.

But, deary me, this is *not* a cookery book.

Pears of some sorts are storable in the same way as apples, but they are dark horses. I always look on them as a crop for the expert: the ordinary mortal cannot possibly tell from the outside whether a pear is ripe, unripe, or bad. Since this book is about food-production, and since the pear is not very valuable as food, I will make discretion the better part of valour and retreat from the pear-salient. The truth is, I don't know any reliable advice to give to the novice, except that some of the most-praised pears, such as Doyenne du Comice, are vastly inferior to the smaller, greener, older varieties.

Brussels sprouts may have the side leaves removed. From now on, decaying outside leaves of these and other greens may be removed and put into the manure pit. And for the rest of the autumn work, *da capo* from the moment when Mrs. B. started to clear the bishop-weed.

FERN OR BULB TROWEL

III

WEEDS AND PESTS

WEEDS have figured largely in this book already. They figure largely in all gardens. I have yet to discover

GROUNDSEL

Hoe before it seeds

whether heavy soil or light is worse for weeds: I think light; my garden is of very light soil.

Against certain weeds, the hoe is a great help if used at the right time. There is no sense whatever in hoeing groundsel when the heads have already turned fluffy with seed; one sows as much as one hoes. The idea of keeping the hoe going all spring and summer is partly to catch the annual weeds before they seed; if one is really successful in this, and keeps it up for a second year and a third, the garden will show results. The yellow flower of the groundsel is the last warning. As to that abomination, shepherd's purse, it must be taken very early, as its lower seed-pods are ready to split and shower out hundreds of seeds while the upper part is still flowering. It makes a derisive noise when it splits.

The continual use of the hoe, or, more laboriously, the small hand-fork, is the only way to deal with annual weeds; seek not out this weed or that, but keep going over all the ground all the time.

If the work is done in dry hot weather, trouble is saved, for the weeds will die on the spot and need not be cleared away. But if the weather is damp, the weeds must be most carefully raked off, or they will grow again. They grow far more easily than flowers or vegetables, and will cheerfully perk up their heads in a shower of rain, although uprooted.

The hand-fork is a more thorough tool than the hoe, which works well enough in rows but is clumsy in working round individual plants. Gardeners (paid) have a rooted objection to the hand-fork, I suppose because it involves stooping; for my part, I don't feel efficient without it.

For the perennial weed, the large fork. It must be dug up and cremated.

The worst perennials to deal with are runners: bishop-weed, the bindweeds, nettles if well established, sheep-sorrel, spear-grass or couch. After trying to deal with these, one feels quite friendly to the thistle and the

dock, which, after all, when dug up are done for. The
runners are never done for as long as your garden has
hedges or a wall; their roots twine in among the hedge-

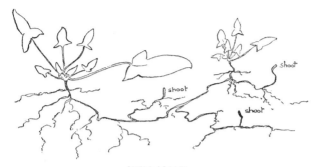

SHEEP-SORREL

Fork it up and burn it! Then treat the soil with lime, chalk or
marl, as it is a sign of acidity

roots, and go right under the wall, so that short of
destroying these amenities you can never totally destroy
the weeds. But on an unfenced allotment, they can be

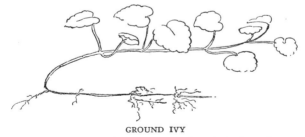

GROUND IVY

Near walls. Pull up. Not a really criminal weed

got rid of by patient forking-up, and if the ground is
well sweetened thereafter by chalk or lime, they will
not care to come again in great force.

"Why bishop-weed?" asks Mrs. B. "I've more often heard it called ground-elder."

"It's the north-country name," says the weather-beaten lady. "There's a lot of Calvinism up there, and they call it bishop-weed, they say, because it grows rank and gross, and strangles other plants."

The B.s have an appetite for slaughter when bishop-

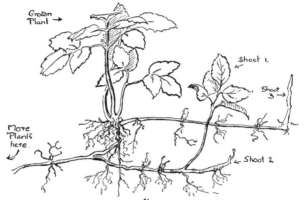

BISHOP-WEED, *alias* GROUND-ELDER
Showing root system. *Fork it up and burn it!*

weed is in question. They go on and on, till dark, with the light of battle in their eyes; there is always a lurking white root spying out at them from the place where they meant to leave off. A fascinating sport, hunting the bishop-weed; in soil of a good crumbly tilth, one can pull out a yard at a time.

Bindweed goes deeper and is more difficult. The spoils are not so large. But it must be tackled; it can do hideous damage if it gets among tall crops such as peas, and ties them all up in knots.

Spear-grass or couch is recognisable by roots striped

neatly in brown and white. Not much shows above
ground—a straggle or two of weak-looking grass. But
underground is a big knot of roots, with runners. It eats
up the nourishment which is needed for one's vegetables.

Sheep-sorrel has such threadlike roots (from which
shoots continually spring) that one could never extract

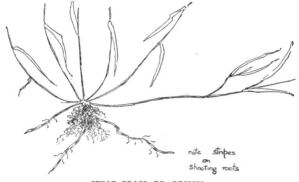

note stripes
on
Shooting roots

SPEAR-GRASS OR COUCH
Fork it up and burn it!

them all; but it is much discouraged by a dressing of
lime, after one has done one's best with the fork.

The mature nettle has roots which run far; like the
bishop, it may be rooted in a quite inaccessible place.
But it can be killed by being cut down three times a
year and never allowed to flower. It is said that the
bishop likewise is sensitive to injury, and may be greatly
reduced if the leaves are pulled off whenever and
wherever they are seen.

Weeds on paths are a different proposition. A heavy
dressing of agricultural salt in spring is a good idea: or
there are plenty of weed-killers on the market which
give "directions on the packet." I prefer salt: I don't

want any dirty looks if a member of my household dies a lingering death. All weed-killers, including salt, must be carefully kept away from plants which are not weeds. Dry weather is suitable for the application of chemicals, dry or showery (but not very wet) will do for salt.

Now for pests. Some have been mentioned already,

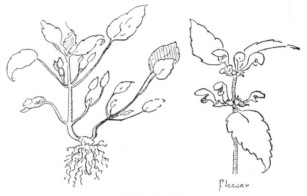

DEAD NETTLE
A hoe will do for this *before* it seeds

but there is an advantage, I think, in having them all dealt with in one place. "All," say I, but human nature is fallible, and this chapter—indeed, the whole book— ought to be foot-noted "E. and O.E."

I am told by Londoners that suburban gardens suffer from slugs and sparrows. One can understand the sparrows, but I should have thought that slugs would avoid such a sooty soil. Both these London afflictions afflict country gardens as well. For slugs, I use soot where possible, putting trails of it alongside but not touching my rows of young plants. It happens to be good for the plants. But it is not a sure preventive in

wet weather, and of late the Meta-fuel remedy has been
much used. One buys a block of Meta at one's usual
shop, breaks it up small and mixes it with bran. It must
be laid well away from plants. It is a pleasanter method
of mass-murder than that of personal assault, in which
the grim-faced gardener seizes the slugs one by one and

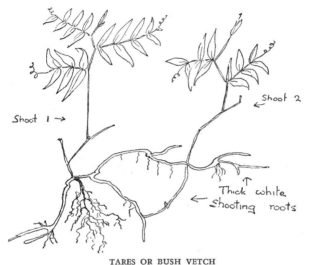

Shoot 1 →

Shoot 2

↑
Thick white
← shooting roots

TARES OR BUSH VETCH

Fork them up and burn them! Then treat the soil with lime,
chalk or marl, as they are a sign of acidity

drops them into a vessel of salt, where they sizzle to
death. Revolting, but efficient, this method.

In fine and dry weather, slugs will not trouble us. I
believe they retire to fastnesses in the lower reaches of
hedges; it is therefore advisable to clear away weeds,
leaves and long grass from the feet of hedges and walls.

The other London curse, the sparrow, is more
attracted by bright flower-petals than by vegetables,

crocuses and plants of the primrose family being his especial prey. They do not concern us here. But the London sparrow may come to learn the unaccustomed pleasure of consuming green vegetables, if many London gardeners become minded to food-production. The only way I know to keep them off is to train black sewing-cotton back and forth on small sticks, two or three inches above the seedlings. It entangles their feet. If bits of white paper are threaded on, it will also, perhaps, scare them. I do not recommend the keeping of a cat in a town garden; I would sooner have sparrows than a cat in my garden any day, much as I love cats.

The sparrow is less of a garden pest in the country (where he is a field pest instead) than certain other birds. Bullfinches bite out the buds of small fruit, blackbirds and thrushes are wretches, no less, and as for jays, if one's garden lies near a wood one can expect only half a pea-crop. Devices against birds include nets and various scaring methods of greater or less effectiveness, but the final remedy is to shoot them and hang a few corpses in conspicuous situations, subject to the close season and the provisions of the protection acts and supplemental county lists. The trouble about birds is that they are so sweet and charming that one feels a brute in warring against them; but in gardening, as in farming, sentiment is out of place. One has to effect a compromise with conscience, and adore the song of the darkling thrush while planning to put a stop to his depredations. In a bushy garden, or a place much wooded and hedged, nesting-time is when the contest with birds may be most effectually waged. But do not take all the eggs—the bird merely makes another nest. Be more subtle: take a pin with you, and *prick* the eggs, putting them

29 CLUB ROOT IN CABBAGE

30 TURNIP JACK OR FLEA (MAGNIFIED)

31 BLACK FLY ON BEAN

32 THE DOCK

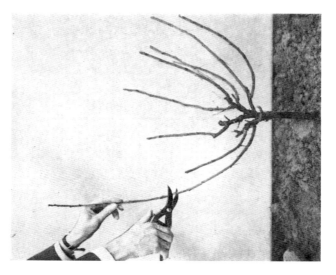

34 HOW TO PRUNE RED CURRANT

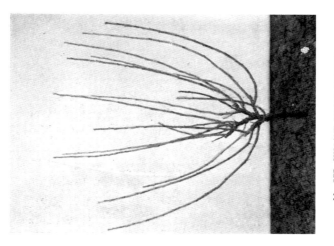

33 RED CURRANT BEFORE PRUNING

back as though they had not been disturbed. The bird then sits out its sitting-time on bad eggs, and when at last it despairs of hatching them, it is perhaps too late in the season to make another nest.

Mice may eat peas and beans in the ground. "Little Nipper" traps baited with something more smelly than peas and beans may lure them to destruction, but it is a fact that garden mice are difficult to deal with, because they are not omnivorous like house mice, and are quite likely to prefer your garden-seeds to even the rankest cheese. There are proprietary brands of poison on the market, designed for rats and mice and warranted non-injurious to other creatures; one of these may be used according to the directions on the packet, for a bad infestation of rodents: but actually things are not often as bad as that. (I have such a tenderness for mice that I hate to think of them dying horribly and slowly. But enough: sentiment is out of place in gardening.)

Moles make clearly defined runs between mole-heaps. A mole-trap may be carefully inserted in a newly-made run, but the chances of catching your mole are remote; in the country, skilled mole-catchers will deal with bad cases. Moles migrate; after a season of dealing ruin, they will depart for someone else's ground.

A solitary mole, though, is no very great matter; it is when they come not in single spies but in battalions that you need professional assistance.

Ants may be a plague on light land; they are not the useful little members of society that we had been led to believe. They hate paraffin, and if the chief holes by which they emerge from their underground colonies are not too near one's plants, one can pour in neat paraffin. If plants are actually attacked, diluted paraffin, in the proportion of a wineglassful to 2 gallons of warm water, mixed with the syringe till it has a milky look, may be sprayed on.

Most of the underground pests may be much discouraged by the application of lime to the ground at the time when it is being prepared for cropping. As much as four or five bushels to a quarter of an acre may be put on in autumn and forked in after lying exposed for three weeks or so. But a proprietary preparation called Vaporite is the only cure I know of for wireworm, a creature whose presence is not detected till it has done its evil work. If young plants turn spindly and yellow, a good many together in a row, that is probably the work of wireworm, and it is too late to do anything about it until next season.

The turnip jack or flea beetle has a way of spoiling turnip and other plants in the cotyledon stage soon after sowing. Soot which has been kept in a heap for some time may be scattered directly on the plants without hurting them, and will put the jack off his food.

Caterpillars should be squashed. As for green and black fly, by far the best method is that of the finger-and-thumb. But most people prefer the less messy one of syringing with some form of soapy wash. Quassia wash is good: boil 1 lb. of quassia chips with 1 lb. of soft soap and a wineglassful of paraffin in a couple of gallons of water; when cold, add double the amount of water, and use with vigour. In all washes applied by syringe, the vigour is half the battle. My own Clensel remedy is easier to prepare, and quite good. By the way, to pick out and destroy the top shoots of the broad-bean plants will help to keep pests down; those are the shoots that tempt the pests, and they are not of use in producing beans.

A general rule in dealing with pests is to remember that poison may poison the wrong person, and that soap is harmless (in the right proportions) to plant-life and human life. If used too strong, it will cause leaf-burning in dry weather. If used without sufficient force, it might just as well not be used at all.

SWAN-NECK HOE,
THE MOST COMFORTABLE
TO USE

IV

THE VEXED QUESTION OF SMALL FRUIT

MOST people embarking on a garden breathe "Strawberries!" in an enraptured undertone, and visualise themselves browsing on their own strawberry-bed almost all the year round. Mr. and Mrs. B. (whom we have not visited for some time) are no exception, and it is with the greatest regret that they come to the conclusion that a quarter of an acre is not enough ground for small fruit *and the rest*. If the choice lies between small fruit and vegetables, and if the country is at war, the vegetables have it. Small fruit, however delicious, is not an essential of good diet.

"No *strawberries*, darling?" wails Mrs. B.

"No room for them. And they're abominably acid things, frightfully bad for rheumatism," says Mr. B. firmly.

But the weather-beaten lady has a large garden, with an established bed of Royal Sovereigns. She says she prefers Paxtons, but finds that for some reason a good stock is less easy to procure from nurserymen than Sovereigns. Mrs. B. yearningly asks her about cultivation, and learns that plants are best taken up and cast away after three years' fruiting, the loss being made good from their runners, planted in early autumn (but runners from first-year plants should be removed as soon as they appear, for the plant's sake, and only runners from older plants saved to make up the bed). The bed should be dug deeply and well manured, as early as possible in the season, so that it is well settled before planting takes place. Plants should be set 18 inches apart, with 30 inches from row to row.

"In spring," says the weather-beaten lady, "you hoe, and you ought also to go out with your thistling-knife; there is a special affinity between strawberries and thistles."

"When you see the green strawberries are grown to the size of an alley-taw—did you ever play marbles?—" says the weather-beaten lady, "you scrounge some wheat straw from somewhere, and lay a handful of it carefully round each plant, under the berries, to keep them from being splashed with earth if it rains. At the first sign of colour, it is time to net them. Lucky people have fruit-cages of wire-netting in which they can walk about upright. I myself put fish-net over hooped hazel-wands cut out of my wood, which is good enough as a protection against the birds, but the very deuce for my poor back and knees when I want to pick the strawberries. It is, however, the cheapest way of doing."

"When the fruit is done," says the weather-beaten lady, "take the straw away, and do some more hoeing. Leave the bed clean before the winter. I often wish I had no strawberries, what with the birds and the slugs and the toads and the squirrels and the hedgehogs, and the weeding and the strawing."

"You console me a little," says Mrs. B. "What about the other small fruits?"

"Raspberries—the secret of success with them is pruning. Keep them down to three new shoots per plant. You plant them in autumn (they like a heavyish soil and lots of muck) and cut off all the fruited canes every autumn. Leave three young canes of last spring's growth to be the next fruiters, and keep them down to three. It seems drastic, but it pays. You propagate them by suckers, of which there are always plenty. Lloyd George is the best raspberry. Some people like autumn fruiters, but I like a summer fruiter like Lloyd George, which fruits in the autumn on the quiet.

"As for currants," says the weather-beaten lady, "it is annoying, but reds and blacks like quite different treatment. Blacks, now, like a moist soil. They are propagated in autumn by cuttings, taken about nine inches long from new growth. There is no mystery about taking cuttings. Make a slanting cut below a bud; rub no buds off, in the case of black currants—for you want a bush, not a standard—and plant your cutting so that there is as much below the ground as above. Cut the top off then, leaving about three buds above ground. It may grow or it may not, according to the weather: a nice spell of delicate showers is what you need.

"To prune blacks, you just thin them for light and air. As to variety, Boskoop is a good one, large-fruited and fairly early.

"But for reds, you choose a lightish, dry soil; and rub the lower buds off your cuttings, because you want a single stem at the bottom. When pruning, you shorten severely all the side shoots growing from the half-dozen main boughs. I don't know what variety to recommend; reds seem to be less of a commercial article, and therefore less 'improved,' than blacks. I have heard of Red Dutch as a good one, and Laxton's Perfection.

"But," perorates the weather-beaten lady, "you won't attempt to grow fruit in your small plot, will you?

You'll be sorry when the fruit-season's over and you have all those bushes taking up valuable ground."

"That's what my husband says," admits Mrs. B. with regret.

"And unless you put the whole plot in with fruit," says the W.L., "you won't touch the fringe of your fruit-consumption. A couple of bushes," says the W.L., "six raspberry canes and one row of strawberries get you nowhere."

"Shall I be able to buy fruit hereabouts?" asks Mrs. B.

"My dear, I'll be glad to sell you some," says the W.L., as warmly as though she had said "give" instead of "sell."

"That's *extremely* kind of you," says Mrs. B., and afterwards wonders why.

DOUBLE CULTIVATOR HOE,
A USEFUL TOOL

V

ARRANGEMENT AND ROTATION

In a garden, as on a farm, there must be rotation of crops, or the ground will grow sick. "Tap-roots," says the weather-beaten lady in her usual dogmatic manner, "should alternate with fibrous-roots: and of course there is the golden rule that *brassica* should never follow *brassica*."

"Of course," says Mrs. B., hypnotised. But her husband shows more spirit.

"It's all very well for you," he says, "you've probably known tap-roots to speak to all your life. But I've never seen a root in the least like a tap."

"Do you mean to say you don't——" begins the W.L.

"Do *you*," he interrupts, "know how to keep books by double entry? Or analyse costs? Or anything like that? I shall enjoy teaching you some day." There is intense meaning behind the words.

But nice little Mrs. B. is both meeker and better-trained to social usages. "Do, please, explain about tap-roots," she begs their rather terrifying friend.

"A tap-root is a single long root such as a carrot—which sometimes gets finger-and-toe disease and doesn't succeed in becoming a single long root. Fibrous-roots are thin roots in bunches, like almost everything else except the tubers: which are potatoes and artichokes, and you have wisely decided not to bother with them. The *brassica* family are cabbages and sprouts and all those things with large green leaves."

"It's going to be very complicated," murmurs Mrs. B., "to remember exactly where everything was, when each plot or row is really so small. I shall make a plan. I shall make a plan for each year."

The W.L. smiles tolerantly, and the B.s feel that she is so full of superior knowledge that she could probably tell by scent where each crop had been. "You amateurs," she says, "always make plans and calendars. But you're perfectly right," she adds suddenly and surprisingly, "and we old hands would do a lot better if we didn't think ourselves above such mechanical aids."

Between them, Mr. and Mrs. B. and their friend plot out their quarter-acre patch. There are 40 rods in a quarter of an acre.

"The cottager," says the W.L., "who always grows potatoes, reckons to plant half his garden with them and to divide the other half roughly into tap-roots and fibrous-roots; then the following year, the half that was potatoes is divided into taps-and-fibres, and the potatoes take up the whole of the former taps-and-fibres part. If you mean to ignore potatoes, you must make up your mind which of the other vegetables you consider most important: let's make a list. Top of the list— greens for the winter, without a shadow of doubt. Those are sprouts, savoy cabbages, broccoli. Men may come and men may go, but sprouts go on for ever; that's how one feels after three months of them. Very useful things, sprouts."

The next most useful things, they thought, were of the bean and pea tribe; then they had an argument as to whether onions were more important than cabbages, and onions won through the strong insistence of Mr. B.

"But the children won't *touch* them," protests his wife.

"They won't touch cabbages either," says Mr. B., "nor eighty-five per cent of our other products."

"I would put carrots before cabbages, too, for usefulness in cooking," says the W.L. "And the least important vegetables are undoubtedly the cucumber and the artichoke, so if there's shortage of space, we leave them out. The leek, too, is not strictly necessary if one has onions, but in a garden of a quarter of an acre one can certainly spare space for leeks. Parsnips, now, are of such a very decided flavour that those who are not with them are very definitely against them; no use growing many. But cauliflowers are liked by nearly everyone. Marrows, wonderfully useful things, we don't bother about because they just go where last year's muck heap was. Now *can* we get some sort of proportion into this?"

Mr. B., luckily, can do sums.

"We have thirteen useful vegetables to put, in varying proportions, into a plot of forty rods, the shape of which is so nearly square that we might call one side 34 yards long and the other 36. Would a row of vegetables take up a yard in width?"

"Taking one with another, I should say rather less," says the W.L., who has become pianissimo now that mathematics are in question. She plots her own garden by rule of thumb, but her own garden is not at all like that of the B.s in size or shape. She feels lost. "Broad beans, 2 foot apart," she murmurs, "runners at least

36 PLANTING POTATOES

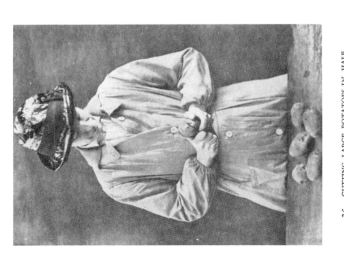

35 CUTTING LARGE POTATOES IN HALF
BETWEEN THE EYES: BOTH HALVES WILL GROW

37   POTATO CLAMP

38   SPROUTING POTATOES: THE RIGHT-HAND SPROUTS ARE TOO LONG,
HAVING HAD INSUFFICIENT LIGHT AND AIR

2 yards, carrots 1 foot, savoys 18 inches, sprouts a yard
—dear me, it's very difficult."

"We must strike an average," says Mr. B. briskly.
"Onions, I think, have rows 9 inches apart. All
round, I should say 2 foot to a row would do for
a first reckoning. Now, we have thirteen kinds of
vegetables——"

"Not more?" says his wife.

"Well, sprouts, broccoli and savoys for winter,
runners and broad beans, peas, beetroot, parsnips,
spring cabbage, turnips, celery, onions, carrots, cauliflowers——"

"That's fourteen, and there's lettuces, radishes and
spinach—very important for the children, spinach. And
we did decide to have leeks. Heavens, I'd no idea so
many vegetables existed. And we're deliberately leaving
out the fancy ones."

"Eighteen," said Mr. B. gloomily.

"Oh, but eighteen's nice. The long side of the garden
is 36 yards, that's 2 yards per vegetable, or three
rows."

"So it is," admits Mr. B. not too generously. "Do
you want me to go on doing this, or will you do it
yourself?"

"Oh, darling, I never could. It was just a little
feminine intuition of mine, about eighteen and thirty-
six. I'm not clever with figures like you."

The weather-beaten lady, who would not fawn on a
man to save her life, can be heard to snort. But it is
one way of getting Mr. B. to do a really annoying job.
If only all vegetables took the same amount of room,
and if only a household would consume the same
amount of every vegetable. . . . So sighs Mr. B. But

once it has been established that two rows of runner beans and three of sprouts can be accommodated if the less popular parsnip and leek, and the less necessary beetroot, are cut down to one row each, with other minor adjustments, life looks brighter, and all that remains to do is to make the plan. As far as possible, root-crops are kept together, also the *brassica* family and the leguminous or podded crops, so that the B.s will know where not to put them the following year.

The weather-beaten lady is rather subdued. Those whom she set out to benefit by her advice have managed, for the most part, without her. But suddenly Mr. B. calls her in conference again.

"I say, what about seeds? Does one buy an ounce, or a pound, or what?"

Here again the lifelong gardener has no ready-made answer. "Depends on the plant. With lettuces, you hope to get about a thousand plants from $\frac{1}{2}$ oz. of seed. With broad beans, you sow a quart to 26 or 27 yards, runner beans a pint to the same distance. With sprouts, you should get five or six hundred plants from $\frac{1}{2}$ oz., cabbages are much the same, so are leeks."

"Gosh, five hundred leeks," mutters Mr. B.

"You'd need $1\frac{1}{2}$ oz. of onion seed, I should say, for three of your rows. I should make it 2 oz. and grow four rows," pursues the expert. "For your one row of parsnips, 1 oz. One oz. of radish seed ought to see you through the season. As to peas, a quart to 40 yards or so."

"But, good heavens," cries Mr. B. in sheer delight, "how jolly cheap it is to grow your own vegetables, compared to buying them in shops! My wife has never bought a hearted lettuce in London for less than three-pence, have you, darling?"

"That," says the weather-beaten lady, "is one of the mysteries of life."

TROWEL

VI

THE HEN AND THE PIG

THE B.s have some rough grass surrounding their cultivated plot. Their whole demesne is triangular in shape,
with the country by-road forming the base of the
triangle, towards which their house faces. The vegetable
plot is behind the house, and behind the plot is the apex
of the triangle, a tufty piece of grass. Alongside the plot
is more grass, adorned with poles and lines, on which
the B.s nurse-housemaid hangs out the nappies of the
infant B.

The triangle is enclosed by overgrown hedges of
thorn, hazel and holly. The angle of the apex is full of
nettles and tin cans. But taking the more or less open
grass-land all in all, there must be forty or fifty rods. It
seems to the B.s a pity to waste it all on nappies. But
they do not want to dig it up and turn it into more
garden, for their existing forty rods, or quarter-acre,
will be more than enough, in a normal season, to supply
vegetables for their family of seven persons.

"Everybody in the country," says Mrs. B., "keeps
chickens."

"Not chickens: hens," says Mr. B., who has discussed
this very matter with the gardener at the Hall and so
has already been laughed at for saying that chickens lay
eggs. Nearly every feathered thing is a chicken to the
town-dweller.

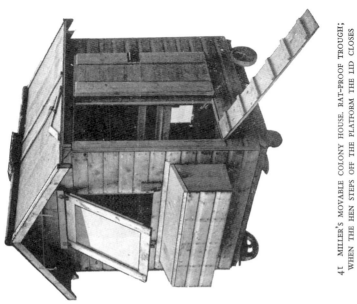

41 MILLER'S MOVABLE COLONY HOUSE. RAT-PROOF TROUGH; WHEN THE HEN STEPS OFF THE PLATFORM THE LID CLOSES

39 WATCHING THEM FEED

40 RAISED FOOD TROUGH

42   SPROUTS AS THEY SHOULD BE   43   A FINE TOMATO PLANT

44   ONIONS LAID OVER BEFORE HARVESTING

"Seriously," says Mr. B., "I've been thinking about it. They eat the household scraps, it appears."

"I," says Mrs. B., "have been thinking about a pig. This rationing . . . 4 oz. of bacon is less than four rashers. Pigs eat household scraps too, don't they? Doesn't one read of the good old days when every cottager had his pig?"

They call the weather-beaten lady into council, and the first thing she says is, "How many household scraps do you suppose you're going to have? Remember, you've *one* destination for them already—the compost pit."

The faces of Mr. and Mrs. B. fall. But they are determined not to waste their few yards of rough grass; far more determined than any country person would be.

Their friend relents and says that actually there is a good deal in favour of keeping a few fowls; in fact, she herself will be glad to let them have some pullets guaranteed to lay throughout the winter for a mere consideration of seven-and-six apiece.

They survey the demesne.

"If you had an old shed," says the W.L., "you would be able to save money. Hens on a non-commercial scale can live practically anywhere that is watertight, ventilated and cleanable. You would have to see that the floor of the shed was firm—be it made of bricks, wood or stamped earth—because it would have to be kept clean by thorough scraping from time to time. Then, if there is a glazed window, replace the glass by wire-netting (or if there is no window, take a piece out of the door), and fit a shutter on a hinge, to be shut at night. Cut a little hole at the bottom of the door large enough for the hens to go in and out: this also should be shuttable at night in case of cats or foxes. Fit perches across your shed, of wood about two-by-three with the edges rounded off, allowing about 12 inches of perch-room to each bird. Down by the wall in a corner set about three or four nest-boxes, our old and tried friend the margarine-box will do, made comfortable and inviting with hay. Then lime-wash the whole place, adding paraffin to combat the red mite, and not for-getting the nest-boxes; and bring in your hens. But unluckily you have no old shed. You will have to go to the expense of a hen-house."

Mr. and Mrs. B. feel like saying, "Hang the expense," but realise that too great a readiness to chuck money about is the mark of the townee who is only playing at country life.

"If you can't make a hen-house yourself," says the W.L., with a sceptical glance at Mr. B., who, although he fancies himself as a handyman on the strength of being able to mend an electric fuse, has indeed never made a hen-house, "I can tell you of a man in the village. But don't be had. He must use something better than half-inch wood; say three-quarters, at all events for the walls; and if it was *my* hen-house, I'd want it weather-boarded—you know, horizontal boards over-lapping to shed the rain. It means longer life to the house than what they call groove-and-tongue fitting. Now, how many hens will you keep?"

They have not the slightest notion.

"How many eggs do you eat? Two dozen a week? More?"

They have not been accustomed, in London, with the price of eggs what it is, to use more than two dozen at most.

"If you keep a dozen hens, you ought to be able to do yourselves proud," says the W.L. "A reasonably good young bird will average nearer three eggs a week than two. Say a dozen——"

"And a cock?"

"And no cock. Not till you know something about the job, if then. By far your easiest plan would be to buy your pullets in early autumn, let them lay as long as they will (they tail off in summer), then eat them and buy some more."

"Buy every year?"

"Well, yes; they fall off in egg-production in their second year, and you don't have the benefit of them as table-birds in the 'chicken' stage. Not that they are chickens in the commercial sense in their second autumn, but if you have any doubts as to their tenderness, stew them very gently first in a little water, and roast them after, basting them well. Older fowls, of course, have to be really boiled, and even if you do roast them after, they taste a bit henny."

"But it seems very extravagant to buy pullets every autumn."

"It's less extravagant for the amateur than making all sorts of mistakes in chick-rearing. Seriously, it is the method I would recommend to beginners. Someone else has had all the risk of disease in the young chickens, and all the expense of the non-productive period, and you step in and reap a practically fool-proof harvest. At a price, naturally. You must pay for laying pullets. But I should call it money well spent. If you yearn for fluffy yellow chicks, my dear woman, it's only excess maternal instinct, and I advise another baby instead. Meantime, what I really needed to know was the number of hens you wanted to keep. If we say a dozen, we can return to the point of departure, which was the poultry-house. Six foot by four or five."

"Not more?"

"They like a good fug. And say four nest-boxes—they don't mind a queue. In a specially-built house the nicest way for nest-boxes is to stick out at the side of the house. They have a lid, which enables you to collect the eggs and keep the boxes clean from the outside.

"But the house itself has to be kept scrupulously clean too, and in a small house this is less easy than in a large one—a man does not have room to stand up in it. Much can be done from the doorway with a muck-rake, so the door should be as big as the end wall allows.

"House-proud hen-keepers have a neat little way with droppings: observing that the worst of the droppings are beneath the perches, they put removable long trays or boards there. We may as well understand here and now that by far the worst part of the hen-keeping is the weekly clean-out. Then, twice a year, there is the internal lime-wash, and once a year the external creo-soting. Your roof, by the way, can be of thinner wood than the walls, covered by tarred felt; and if you have a gutter to catch rain-water, leading to a tub, you will be saving yourself some trouble. The eaves of the roof should project well over the edges of the walls.

"That top corner of your piece of grass is just the place for the house: it will face south for warmth, and yet have the advantage of the overhanging hedges to protect against excessive warmth. Don't let the hedges overhang too much; their tallness is an advantage, but their drips aren't."

The W.L. pauses for breath and thinks of something

else. "If you were really pressed for money, you could have a lean-to against that coal-shed of yours by the back-door. Not so much sun there, but a westerly aspect is next-best to a southerly. You would be sure to run the rain-water from the roof into some sort of receptacle, to keep the backyard dry. I have seen happy hens in a yard. And then that piece of grass would do for the pig. . . . But let me finish with the generalities of hen-keeping first.

"You want winter layers which will have nice breasts on them when you come to eat them: I suggest Rhode Island Reds or White Wyandottes. There are breeds which lay more, such as Leghorns, but the heavier fowls are both better winter layers and nicer to eat. Their trouble is, they will sit. Unless you let them raise chickens, you will have to have a little penitentiary made, a sort of coop on legs, with a slatted floor, to make them sorry they thought of sitting.

"Put a high fence of wire-netting round as much of the grass-land as you mean to let them have. It won't be grass-land for very long—it will be just a trodden patch of earth. They eat the grass. If you had more land, you would change their grass run from time to time, but it is not necessary if you see that they get exercise in their restricted one. Hang shot lettuces or occasional cabbages up on the netting, where they have to jump for them. Scatter a little crushed oyster-shell and road-grit now and then—they need grit in their crop, and it is good to make them scratch about for it. When they have stamped-down their piece of grass, it may become very foul and wet; it is a good idea to divide the run into three, and floor the middle part of it with well-firmed gravel, leaving the other two in grass, then let your hens have access to the gravel and one grass-plot while the other grass-plot rests and grows green. That is, use the grass-plots turn about, and the gravel all the time; it does not matter that the plots will be very tiny.

"About food: your household scraps, of which we have already heard so much, will be the most valuable part of the diet, and remember that the hen would be a meat-eater if she had the chance, so give her the plate-scrapings of bits of meat, fat and gristle, and fish-scraps too, but not too much. All vegetable scraps are good, but better cooked than uncooked—and *not* potato-peelings except if they are cooked separately and have all the water poured off. They may then be mixed with biscuit-meal to a crumbly texture. No food should be given sloppy except for fattening purposes; the crumbiness is the indication of the right consistency for soft food. Keep a special saucepan for cooking the hen-food, and cook it every day or at most every two days, adding sharps or middlings (bought from the grist-miller or the corn-merchant in your market-town) to bring the stew to its crumbly state after cooking. Give it to the hens hot, in winter-time, in the evenings—and if you have skim or sour milk to mix in, so much the better. The morning meal is of grain—barley, oats, wheat and maize, but with the wheat and maize used sparingly. Whatever you are feeding to them, give them a little at a time and see that they clear it up; it is a slow job, the first few times, but then you get to know roughly what they will want. To leave surplus food in the run is to make a mess and to encourage rats and mice and such small deer.

"The grain feed should be scattered well about, and if it is given in a trodden-down yard, some dry bracken

or other litter such as chaffed wheat straw spread about
the yard will partly hide the grain and compel the hens
to scratch for it. You understand that exercise is a
problem when birds are kept in a confined space.

"People who keep poultry on a commercial scale find
it pays in the long run to have good equipment and
plenty of it, but the few birds kept to supply the household, by numbers of small householders, do not repay
lavish expenditure, nor do they need anything but a
water-trough and mash-trough. These, of course, must
be kept clean, and fresh water given every day.

"When cleaning out your hen-house and run, once
a week, remember that poultry-manure is very valuable
for the garden. It is the richest natural fertiliser we have,
but must not be used fresh or wet. Store it under a
shelter somewhere, and spread it now and then to dry
it. The litter from the scratching-run will add to its
value in humus and aeration. If the hens stamp down
their litter too much, loosen it with a garden fork: they
won't scratch if it looks hard and lumpy."

Mr. and Mrs. B. are somewhat awed at their friend's
voluminous knowledge, but she hastens to disabuse
them: "What I've told you is only the bones of the
matter. Myself, I do everything for the easiest. But you
can complicate your life enormously if you care to, by
buying books on the subject and doing everything just
right. I knew a woman with a baby once who weighed
the poor little devil before and after each feed; what
she did to it if she found it had eaten too much I never
discovered! I admit that it grew up to be a fine child: I
have never denied that the experts are usually right. But
the inexperts don't go so far wrong either if they use
common sense; and life is short. . . . You are at liberty

to weigh out the exactly correct amount of food for a pen of twelve laying pullets, but you'll come to much the same result if you feed them by handfuls and see that they leave a clean plate."

"What about diseases and pests?" inquires Mrs. B., who has already learnt, from her garden, that nothing is immune from them.

"Red mite is the commonest; you'll keep it down by thoroughly lime-washing, as I've said, and adding paraffin to the wash. See that every crevice is filled, and look to such places as the ends of perches. Scaly-leg is caused by another parasite: if you notice the legs of your hens becoming rough and thickened, soak them in paraffin for a few minutes before bedtime, and this will bring the scaly part right off. Burn it. For any apparent upset of the bowels, give the hen a teaspoonful of warm salad oil. If you see a bird moping about with a full crop, she may be crop-bound: give her frequent doses of warm water and bicarbonate of soda, also a dose of salad oil, and massage the crop. She might need a surgical operation, which you could quite easily do, if you remember to sterilise the scissors and needle——"

Mrs. B. shudders.

Unmoved, the W.L. continues, "Open the crop, take out the contents, wash it out with tepid water, and sew it up again. Starve the patient for a day, and then feed on bread and milk. But the complaint is not very common."

"Thank heaven for that. Imagine operating on a hen."

"And as for real diseases, you avoid the worst of them, such as B.W.D. and coccidiosis, by buying grown pullets instead of raising chicks. Healthy-looking pullets

with large, bright eyes, well-coloured beaks and legs,
and an alert manner of scratching for food, should
remain healthy until you want to eat them, if they have
a sunny run, are not overcrowded, nor overfed, and
are guarded against parasites in the ways I have mentioned. They don't even moult in their first year. If a
bird seems ill and is not amenable to treatment for
simple bowel or crop trouble, take it away and slay it,
but don't eat it—it may be tubercular."

"How does one slay it?"

"One goes to an expert for a demonstration. To tell
you is not to instruct you fully. You take the body and
wings firmly in your left hand and arm, and the head
in your right hand with the skull between the first and
second finger, then you pull and twist at the same time:
pull to separate the vertebrae and twist so that they are
completely dislocated. The bird may kick about a bit
after death, but if you have done the job properly, it
has died instantaneously. Keep the head hanging down,
and pluck as soon as may be, in the case of a bird killed
for the table.

"But we've reached the end of the bird's history too
soon: I ought to have mentioned a troublesome vice—
egg-eating. You can detect a Cannibal Ann because she
doesn't use a dinner-napkin, and when you have spotted
her, isolate her and starve her for twenty-four hours,
leaving beside her a blown egg filled with mustard and
red pepper. That will teach her."

So hens, it seemed, were no more of a mystery than
the garden. "Just a little common sense," says Mr. B.
blithely.

"And a little method, routine and determination,"
adds Mrs. B., reflecting that the hens have to be fed

twice a day, for three hundred and sixty-five days in every year, and cleaned out once a week, for every week of the fifty-two.

"But all those things are included in common sense," says Mr. B.

"Would you say that the one thing needful for a successful life was common sense?"

"Yes," says Mr. B., "for a *successful* life. Of course, you want luck as well, if you're to be happy. And something more than that if you're to be great. Vision, would you call it?"

But Mrs. B. is not a philosopher. "My vision," she says, "is of a pig."

"Ambitious, aren't you?"

"Well, but in the good old days everybody in the country had a pig. The cottager's pig was his mainstay. Surely I've read all that."

"And why doesn't the cottager still have a pig? He doesn't, you know. The only pigs round here are on farms."

Mrs. B. did not know the answer to that. As a matter of fact there are several answers, as the B.s found out when they inquired into the matter of pig-keeping.

In the first place, the Sanitary Inspector steps in if pigs are kept too near houses; that has put an end to many a cottager's pigsty. And the law allows neighbours to complain of nuisance on the very slight provocation of a smell of pig when the wind is in a certain quarter. Town people who come to live in the country are always ready to complain of cocks crowing and dogs barking and pigs stinking, though they ought to be well disciplined to nuisances, what with buzzers screeching and cars hooting, and buses and fried-fish shops stinking worse than any pig.

But the real reason for the decline in cottage pig-keeping is a sadder one. It is nothing to do with the Pig Marketing Board, for that much-execrated body only deals with baconers, and the pork-market is still open. It is to do with changed methods of house-keeping. Now that the cooking in the cottage is all done with a tin-opener and a frying-pan, there are few scraps for the pig.

Mr. and Mrs. B. visit a neighbouring farm, just to look at pigs. What they find depresses them somewhat: it all looks so expensive. The farmer's pigs live in clean straw on concrete floors; at one end of the sty is a raised platform, straw-covered, which is the bed. The sty faces south, and is light and airy, with a fenced run adjoining in the open air. "I don't see the central heat-ing or the bath," says Mr. B., intending irony.

"What adorably clean pigs," croons Mrs. B.

Pigs, they are told, would always be clean if they lived in the open as nature intended. But to keep them clean in confinement is not easy. The concrete floors, sloping slightly down towards a gutter which takes off the liquid manure, are almost essential. But as concrete is cold, and pigs are liable to cramp, they must be well littered. Straw litter stamps down into the best manure for the garden, and from the pig's point of view wheat straw is the cleanest and least musty, but there are many demands on a farmer's stock of wheat straw, and it is unlikely that stray pig-keepers such as the B.s would be able to get a large enough supply. Bracken, or heather, they learn, will do well.

"But you realise," says the farmer, "that if you want

muck for your garden, there's going to be a tidy old stink around your pigsty. If you were thinking of keeping a pig as a pet, you could clean him out every day, but if you want him as a muck-machine, you've got to let him make some muck. These pigsties of mine— they're only clean on top. Underneath the clean straw there's 6 inches of something a bit different, and my word, my missus has something to say to me if the wind sets in the west the day we muck 'em out."

Mr. and Mrs. B. are weakening on pigs. "Would a pig live on house scraps?" they ask.

"Why, no."

"But didn't the cottager's pigs live on house-scraps?"

"Not entirely. The cottager was always a farm-hand, who could buy barley-meal and so forth cheaply from his master; generally he got the pig for nothing, too, the pipman of the litter. I should say that not more than one-third of the diet of the fattening-pig would be house-scraps; the rest would be middlings, barley-meal, maize-meal, fish-meal and so on, the staple being the barley. The law says that all 'wash' fed to pigs must be cooked; it's often ignored, but to be on the safe side you should boil your house-scraps for your pigs every two days."

"And when the pig is fat, can we use it for bacon ourselves?"

"They tell me that, owing to the war, regulations are to be loosened a bit, so that folk can kill their own pigs, as long as they use the humane killer, and can eat their own product, 'spite of rationing and all that. But that fare to me a bit of a speculation for the amateur. Young pigs are delicate customers. I reckon you'd pay dear for your bacon, even supposing you could get the fattening foods, which are to be greatly restricted so that the ships can bring munitions instead. A farmer, give him time to grow it, can grow the grain foods for his pigs, but you folks can't. If I were you (and it would be every bit as patriotic or more) I would buy an in-pig sow and sell the litter. She's not so hard to feed as fattening-pigs, and her litter will usually keep healthy so long as they're feeding on their mother. As to the old gal, she needs to be fed differently at different stages. To begin with, to every 100 lb. of her weight (have her weighed when you buy her) she wants a daily ration of 3 lb. of middlings and 1 lb. of bran, mixed with your household 'wash'; then about a month before she farrows she needs half a pound of fish-meal a day in addition. While she is suckling the young 'uns—but don't feed her the first day—start her on several warm feeds a day, of equal parts of middlings and ground oats, with some skim milk in the wash to mix. After four days of this, put her on to a good diet consisting of:

> 3 parts middlings.
> 2 parts barley meal.
> 2 parts maize meal.
> 1 part palm kernel meal.
> 2 parts white fish meal.
> 1 part bran.

That is, supposing you can get all these things. I read every day in my paper letters from farmers asking where their feeding-stuffs are to come from. We've been made to fall in with a wrong policy for years. It's been dinned into us how nice it is to buy cheap foreign feeding-stuffs for our stock and turn more and more of our land to grass at home—and now th'owd Government is reaping what it has sowed. But just for fun we'll *suppose* you can keep a pig."

Mr. and Mrs. B. are all sad attention.

"You mix that ration with your house wash. The young pigs will eat it as soon as they are able. But as they grow, they go on taking more from their mother, so you need to make sure she is having enough. Don't let food go stale in the troughs; it pays to watch at meal-times and make sure the pigs clean up what you give them.

"To wean 'em, start in the eighth week to remove them from their mother, putting them into a separate run for longer periods each day. Cut down the sow's ration to 2 lb. of middlings and ground oats, mixed with wash to a total of 4 lb. a day to each 100 lb. of her weight. When the pigs are weaned, you'll sell them as soon as may be, and meanwhile feed them on equal parts of barley meal, ground oats and middlings mixed with skim milk or house wash. There's another thing—skim milk is the very best food for fattening pigs, but where can the ordinary householder get it? Well, you should also put a heap of wood ash and cinders in a corner of the sty, and a piece of rock salt. If you watch the little beggars feed, your own sense will tell you when to give them more food on account of their growth. In summer, green veges., and in winter, roots may be given as extras, uncooked. But potatoes should be cooked, unless they are only the few 'chats' from your garden."

"How long before the pigs are fit to sell?"

"Sell 'em as weaners, and you'll be saved no end of worry. At eight weeks or soon after. Then someone else has all the bother of erysipelas, swine fever, rheumatism and what-all. Surprisin' delicate things, pigs, but, same as babies, they don't go far amiss while their mother feeds 'em. Moreover, if you let 'em grow big, you'll need far more space for them."

"How much money ought we to get for weaners?"

"Twenty-five bob to forty bob each, according to size and the market. A good proposition."

Mrs. B. has misgivings on another point.

"Is there no special skill needed when the sow farrows?"

"Bless you, no. You do more harm than good, messin' around when she's busy at that job. But I find it a good plan to have a rounded rail, such as a length of iron piping, running round the sty 1 foot away from the walls and 8 inches off the floor, so that the piglings can get away if she is clumsy and wants to lie on them. You see, she'll lie up agin the rail instead. Don't have too big a sty; 10 feet by 12 will do. The run can be the same size, or as large as you like."

Mr. and Mrs. B. think they had better keep out of the pig world. Pigs, clearly, are a more skilled job than poultry. And while poultry will amplify the diet of the B. household, the keeping of a sow in order to sell weaners would be pure altruism. Or no, not quite, the farmer points out: there will be a tangible profit if the old gal raises even as few as eight weaners twice a year.

Avarice stirs in the breast of Mr. B., but his wife is firm. "Not yet," she says. "Not till we see how much wash there would be for a pig, after the poultry have had their whack."

In the event, they find that their quarter-acre of vegetable garden, when fully productive for the needs of their household of seven persons, gives rise to a whole bucket a day of leaves, stalks, peelings and cooking-water. It is more than enough for a dozen hens, but the claims of the compost pit must be met, and it seems that there would be no noteworthy economy in keeping a pig: at all events while grain-feed is so hard to come by.

But what is a country gentleman without a pig? What is Sunday in the country without a pigsty wall to lean upon and a rough pig-spine to scratch with a meditative walking-stick? I feel sure that the B.s will succumb to an in-pig gilt as soon as their hens and their garden are so ordered as to take less of their anxious attention.

VII

A NOTE ON FLOWERS

IT was quite early in their acquaintance that the weather-beaten lady said, "I have some wallflowers over: would you like them?"

Mrs. B. drew herself up. "Oh, I don't think so—it's very kind of you, but, you see, we're going to grow *food.*" Patriotism and virtue exuded from her.

The weatherbeaten lady was not the sort of person to bring out pat, "What about food for the soul?" It was what she would have liked to say, but she was unaccustomed to speaking of souls. Manures and mulches were more in her line. She said eventually, "A garden doesn't seem like a garden without flowers."

Mrs. B. responded in a superior manner, "But in a time like this there will be great changes: much greater than merely turning to vegetables instead of flowers in the garden. We must think of the greatest good of the greatest number."

"And no one is going to think of the good of the poor nurseryman, I suppose," said the weather-beaten lady, "who has stocks of flower-seeds and plants on his hands."

"He'll sell all the more vegetable seeds and plants, surely," said Mrs. B.

"In a large number of cases, the man who sells vegetables is not the same as the man who sells flowers.

Flowers are a specialised job. Nor does it really help anybody to have flowers a drug on the market and vegetables in such great demand that there's a shortage, and consequently an increase in price. There should be moderation in all things, I used to be told when I was amenable to proverbs. And now look here: do you actually dislike spring flowers?"

"Why, no, of course not. No one could dislike spring flowers."

"From your tone I gather that you're fond of them."

"Naturally."

"Well, then, be as much of a glutton as you like over your vegetables—you can still have quantities of bulbs, wallflowers and forget-me-nots. A lot of your piece of garden will lie empty till May or June (the places reserved for your late sowings of successions of vegetables), and by the end of May, don't you see, you yank your spring flowers right out. They don't conflict with your vegetables. Then again, when a vegetable row is definitely done with, unless you have the place earmarked for a late sowing you might as well put in some summer bedders."

Mrs. B. is weakening. "There used to be round beds in front of the house, it seems," she says.

"Oh, roses for those. Nothing but roses," says the W.L. masterfully.

"Isn't rose-growing a deep mystery? We are such beginners——"

"You've got to begin with roses some time, haven't you? My advice is, don't begin later than November. Let's look at the beds."

The weedy round beds proved to be of friable, medium-light soil like the rest of the garden.

"Lovely soil for most things, but not too good for roses," pronounces the W.L. "When you've got it free of weeds and deeply dug and well mucked and are making the large holes for the rose-roots, it wouldn't be a bad idea to put in the bottom of each hole a turf of grass with the grass downwards—to hold moisture— or else a chunk of clay, nearly the size of your head, broken a little and mixed with the soil. You need to dig your hole more than two foot deep, to do it properly.

"You should buy your roses from a nursery near by, or on the same sort of soil, then the change won't be too great: but they will benefit from either the grass turf or the clay in any case. A good rose plant has its root-system shaped in a fairly neat bunch; there should be large stiff roots with small fibrous ones leading from them. If there is an extra long, straggling root, you ought to shorten it with your secateurs. See that the roots have room in the hole, and tread all round the plant to make it firm and upright. Don't prune it till late March or April. Then if it looks weakly, cut it back thoroughly, leaving only four or five buds on each shoot, but if it seems a strong grower, leave it more buds to spend its strength on, say eight to a shoot. Clear your prunings away, and of course take out all dead wood and old hard-looking wood with only spindly new shoots growing from it.

"In a real drought, roses must be watered—given not a sprinkling but a soaking. A mulch of grass clippings round the stem helps to keep them damp, and a mulch of well-rotted horse-manure before flowering-time will help in every way. Look out for greenfly, and on light soil for ants. You will see the ants on the young flower-buds, eating stolidly through from tip to base."

"What kinds of roses . . . ?" suggests Mrs. B. when the weather-beaten lady pauses for breath.

"Old and tried friends, until you know what your soil is really like. Betty Uprichard, for a pinky-orange, Etoile de Hollande for a red, George Dickson, a darker red, Mrs. Sam McGredy, a light yellow. Those will fill your round beds. Then when the war is over you'll have become rose-fiends, and can launch out all you like. Only, it's shockingly expensive, plunging for the latest as soon as it appears."

The W.L., who is in communicative mood, reverts to the spring flowers and explains how she always sows wallflowers in June, and gives them their first trans- plantation in August, and their final bed-out in October or November. Manure is forbidden them. Blood-red, Cloth of Gold and Vulcan are her pet kinds. Forget- me-nots, she says, have their caprices. They will not stand drought; but in a dampish or semi-shaded position they will seed themselves and keep the garden replenished year after year. It is advisable, she says, to get a good variety, such as Sutton's Royal Blue, for a start, then after a year or two, when the seedlings seem to be losing colour a little, buy in some more.

As for bulbs, the most utilitarian garden is no garden without bulbs: nothing is a garden without bulbs: con- versely, a garden is nothing without bulbs. They take up so little room: like cordon apples, they stand up straight and thin, and give maximum value for their space. The B.s decide that where the little path runs through their plot, edged with parsley and herbs and fenced with espalier or lateral-cordon apples, there will be room for bulbs; and round the house, and edging the rose-beds. Indeed, where will there not be room for bulbs?

"Buy cheap daffodils for naturalisation, as the catalogue calls it," advises their mentor, "and naturalise them—get one of those sharp narrow trowels made for the job, and stick them in the turf just anywhere and anyhow. Do the same with crocuses. But tulips, ah, tulips . . . the double earlies are like little flames of light. They need to be in rows or groups, to take the eye. I call to mind Murillo and Tea Rose. The best of the tall later ones is Pride of Haarlem—so red that it makes you gasp. And I've had some lovely groups of the greyish-pearly Painted Lady and Tulipe Noire, together. And there's a salmon-rosy creation called Nova which I can't live without. Those will do you for a start. They ought to go in in late October or early November: daffodils and crocuses earlier."

Mr. and Mrs. B. are by no means sure that they will succumb to the temptation to plant spring flowers and roses. But they will. Man cannot live by vegetables alone. And when they have seen the long lasting and delicately radiant bed of single asters, Southcote variety, which is the pride of the W.L. in August and September (raised from seed in cool greenhouse in March, pricked into boxes in April, planted out in late May) they will probably commit themselves to all the trials and troubles of bedders-out. Flowers are irresistible.

VIII

A B C OF USEFUL VEGETABLES

Artichoke, Globe.—For large gardens only. Plant suckers in spring. Plants may live six years or so.

Artichoke, Tuber.—For large gardens only. Plant in spring, a foot apart, a yard between rows. Grows easily anywhere.

Asparagus.—For large gardens only. Plant Connover's Colossal crowns two or three years old, in sandy soil in April, allowing 18 inches between plants and between rows. Plant 4 inches deep, preferably on long raised beds, say 2 rows to a bed. Keep free of weeds; on light soil, dress with salt in spring, and in autumn cut down the dead stems and spread well-rotted farmyard manure. A perennial plant.

Broad Bean.—Sow Longpods in November, and again in January. Any variety preferred may be sown subsequently in three-weeks succession. Seeds 6 inches apart, rows 2 feet. Hoe well round the plants.

Runner Bean.—Sow May and June, 3 inches deep, 12 inches apart, and leave 2 yards from row to row. Stake well with long stout stakes. Painted Lady is a good sort.

Beetroot.—Dig deep. Sow in April (late) or May. Thin out to 8 or 9 inches apart, with rows 12 inches apart. Beet should be on land manured for the *previous* crop. The globe shape is best.

Broccoli.—Likes heavy soil. Sow March, April or May, thin out to 2 feet apart, with 2 feet between the rows. Do not try to grow the Roscoff varieties, except in the warm south-west of England; for general planting, the kind called Self-protecting is good (see Cauliflower, which is almost the same thing).

Brussels Sprouts.—Thrive best in soil which is not very light. Sow in March, plant out in May, leaving 2 feet between plants and a yard between rows. Make the soil very firm. Cut the sprouts from the bottom first. as soon as they look large enough, and should there be a shortage of greens when the sprouts themselves are done, the middles of the plant-tops are quite eatable. Suttons have a good kind called Matchless.

Cabbage.—Sow in March for spring use, July and August for winter. Do not put a plant of this family in ground where plants of similar nature have been the previous season. Thin the seedlings if they come up crowded, and plant out in September, 18 inches apart, with 2 feet between rows. Varieties for spring: Ellam's Early and Offenham; for early autumn use Primo and Autumn; then comes Christmas Drumhead, followed by

Cabbage, Savoy.—Likes heavyish soil. Sow in March and plant out in June or July in the same distances as other cabbage. Varieties: New Year savoy, and a later one called Sutton's Rearguard, which is ready for use in February and March.

Carrot.—Stump-rooted kinds are more generally successful. Sow in March, April or May and thin well. It pays to buy good seed of this vegetable (i.e. from a well-known firm and probably rather expensive) as carrot-seed varies a great deal. The soil need not be rich or heavy, but should be deeply worked before sowing.

Cauliflower.—Almost the same thing as Broccoli, but generally understood to be used when broccoli are done, i.e. May onwards. They need good manuring and deep digging. Sow in August or September. Snowball is a good kind for eating in June, and Sutton's White Queen for July and after.

Celery.—Sow in boxes in cool greenhouse in March. Plant out when large enough. Some people plant in trenches, and do not earth-in the trenches till autumn; others plant on the flat and earth-up the plants progressively as they grow, beginning when they are about a foot high. Drought affects them badly, so keep watering in dry weather; and when earthing-up, it may be advisable to tie the tops lightly together, to keep the earth out of the middle of the plant. Some growers go so far as to protect the plants with paper tubes. There are many varieties; Cole's Crystal White is good.

Cress.—Sow in boxes in cool greenhouse, keep lightly watered, use in the cotyledon stage—i.e. when there appear to be only two "leaves."

Leek.—Sow in March. Transplant the good-looking seedlings when about 6 inches high into holes made with a dibber, deep enough to hide almost the whole plant. Just drop in enough earth to make the plant secure, and firm it down with a stick: do not fill in the holes till autumn.

Lettuce.—Likes a fairly light soil. Sow in April or May, and in succession through summer as required; sow a winter variety such as Sutton's Imperial, in autumn to stay outdoors all winter and be eaten in spring.

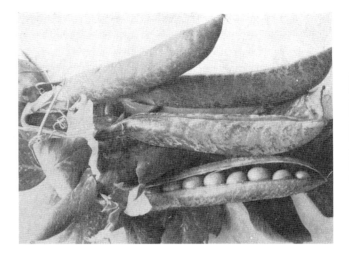

46 PEAS, GOOD FULL PODS

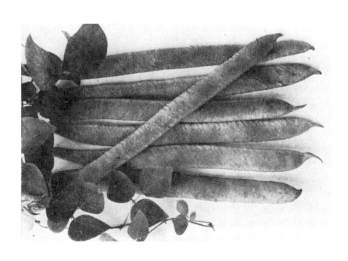

45 RUNNER BEANS

47   CABBAGE LETTUCE

48   SAVOY

Thin out seedlings well, and transplant the thinnings. Do not allow young plants to suffer from drought. All The Year Round is a good spring–summer kind, cabbage-shaped, and for the long or cos variety try Sutton's White Heart. I always tie the tops of my cos lettuce loosely as soon as the outer leaves look capable of shutting-in the hearts; this whitens the heart and discourages "bolting."

Mustard.—Grown like cress.

Onion.—Dig the ground well. Sow in March and April, and again in August. Thin well, to about 6 inches apart, and use the thinnings for salad if liked. To harvest onions, take them up on a dry day, let them lie on sacks in the sun for a few days, taking them in at night: the moving-about will knock any remaining soil from their roots. Then hang up in bunches by the tops, on an airy and dry wall in shed or outhouse. Bedfordshire Champion and Ailsa Craig are the pick of onion varieties.

Parsnip.—Deep digging means long roots. Do not sow on land recently manured; land from which a crop has already been taken will do. Sow in March. Thin out to 9 inches apart. Do not harvest until the plants have tasted a frost. Do not serve at table plain boiled!

Pea.—The queen of vegetables. Sow in March, April and May on generously manured land (in February in sheltered places) 2 inches deep and three or more inches apart. Stake when they seem to need it. The best kinds are: earlies—Thomas Laxton and Improved Pilot; maincrop—Onward and Telegraph. Remember to cook with a teaspoonful of sugar.

Potato.—For large gardens only. Really a farm crop. Sow earlies in the end of March—Arran Pilot or Eclipse.

Early cropping will be obtained if the seed (i.e. small potatoes) has been set to sprout in shallow slatted trays in a place under cover and secure from frost for a month before planting. Do not break the sprouts when planting, but rub off all sprouts above the number of two. Soil should be crumby and well drained, and should have been liberally manured the previous autumn, and perhaps dressed with lime. When leaves appear, begin to earth-up the plants, leaving only a little green exposed; when the earthing-up is finally done, the ridges will be about a foot high.

Maincrop potatoes are: Majestic or King Edward for light or medium soil, Arran Chief or Doon Star for stiffer soil. In all cases, do not keep growing your own seed, but obtain seed from Scotland (or guaranteed to be from Scotland by your seedsman) every other year.

Early potatoes are eaten as they are taken up; maincrops are stored in a "clamp," a long heap—the bottom of which should be on slightly raised earth if the land is not very well drained—covered with straw and then again well covered with earth slabbed-down with the spade, ventilation holes being left here and there at the top, with loose twists of straw in them.

Radish.—Sow from February, every fortnight or three weeks if liked. The turnip shape is the more popular.

Rhubarb.—Propagation by seed is slow; by root-division is better. May be planted in autumn, though many advise February or March. Cover in January with straw and forcing-pot to obtain early stems.

Savoy.—See Cabbage.

Spinach.—Manure the soil well. Sow at intervals from

March onwards, and do not allow to suffer from drought. Thin out a little. "Round" varieties mature quickly, and spinach-beet is a cut-and-come-again kind. *Note.*—Before spinach is ready a delicious imitation may be made from young nettletops, well washed, cooked without water except that which clings to them (spinach itself is cooked this way too), put through a sieve, and served with a little cream stirred in.

Tomato.—A greenhouse crop. Though, in a favourable year, tomatoes may be successfully grown against a sunny wall, they cannot be depended upon. People with cool greenhouses are well advised to buy young plants, as two or three transplantings are needed if tomatoes are grown from seed. Regular watering is necessary, and growers should watch for white fly, which can be controlled by the usual anti-pest washes if taken in time. Side shoots must be removed and the top shoot should be taken out in early August, so that the plant can give all its strength to the fruit. As it is hardly worth while filling one's greenhouse with tomatoes at a time when the fruit is cheap to buy, I advise any variety advertised by a reputable firm as *very early*.

Turnip.—Likes rather sandy soil. Sow at intervals from May onwards, and thin to five or six inches apart. Turnips need to grow quickly—i.e. in damp weather —if they are not to be woody and strong-flavoured, but we cannot control the weather. We can, however, make sure of using our turnips young enough, that is, within eight to ten weeks of sowing. Watch for the turnip-jack, which attacks the plants in the cotyledon stage. Soot discourages him.

Vegetable Marrow.—Sow when frosts are over (May?) or else in the cool greenhouse in March for transplanting. Likes manure for a foot-warmer; soil tramped over the place where an old manure-heap has been is a good bed for it. Keep watered. Old and ripe marrows will last well into the autumn for jam and chutney-making, but as a vegetable they are best eaten young—about 15 weeks after planting.

HAND-FORK

49 LONG-ROOTED PARSNIPS

50 STUMP-ROOTED CARROTS

51 FINE CAULIFLOWER HEADS

52   RADISHES IN ABUNDANCE

INDEX

(The numerals in heavy type refer to the figure-numbers of illustrations)